Invertebrates - Ecophysiology and Management

Edited by Sajal Ray, Genaro Diarte-Plata and Ruth Escamilla-Montes

Published in London, United Kingdom

IntechOpen

Supporting open minds since 2005

Invertebrates - Ecophysiology and Management
http://dx.doi.org/10.5772/intechopen.78745
Edited by Sajal Ray, Genaro Diarte-Plata and Ruth Escamilla-Montes

Contributors
Jeffrey Marliave, Laura Borden, Jessica Schultz, Donna Gibbs, Glen Dennison, Caroline Eve Dubé, Chloé A-F Bourmaud, Emilie Boissin, Alexandre Mercière, Serge Planes, Nandini Ray Chaudhury, Ashwin Gujrati, Tvr Murthy, Chowdula Satyanarayana, José Manuel Mazón-Suástegui, Abel Betanzos-Vega, Mario Formoso-García, Avilés-Quevedo María Araceli, Raphael Abdoulaye Ndione, T. Nhan Hua, Ako Harry, Ricardo Sousa, João Delgado, Joana Vasconcelos, Rodrigo Riera, Ana Rita Pinto, Lídia Gouveia, Adriana Alves, José António González, Mafalda Freitas, Paulo Henriques, George Andrei Draghici, Bindhu Jayaprakash, Arunava Das, Belen Lumeran

First published in London, United Kingdom, 2020 by IntechOpen
IntechOpen is the global imprint of INTECHOPEN LIMITED, registered in England and Wales, registration number: 11086078, 7th floor, 10 Lower Thames Street, London, EC3R 6AF, United Kingdom
Printed in Croatia

British Library Cataloguing-in-Publication Data
A catalogue record for this book is available from the British Library

Additional hard and PDF copies can be obtained from orders@intechopen.com

Invertebrates - Ecophysiology and Management
Edited by Sajal Ray, Genaro Diarte-Plata and Ruth Escamilla-Montes
p. cm.
Print ISBN 978-1-83968-551-4
Online ISBN 978-1-83968-552-1
eBook (PDF) ISBN 978-1-83968-553-8

For EU product safety concerns:
IN TECH d.o.o., Prolaz Marije Krucifikse Kozulić 3, 51000 Rijeka, Croatia, info@intechopen.com or visit our website at intechopen.com.

Meet the editors

Sajal Ray received MSc and MPhil degrees in Zoology and Environmental Science from Calcutta University, and a PhD from Jadavpur University. His thesis reports immunotoxicity of pesticide in an economically important snail of India. As an awardee of Fogarty Visiting Fellowship, Dr. Ray carried out his postdoctoral research in cardiac pathology at the National Institutes of Health, USA. His research interest is immunological responses of molluscs, sponges, crabs and earthworms exposed to pollutants. His team is engaged in understanding the evolutionary mechanism of immunity in phylogeny. He presented his research in conferences including the World Congress of Malacology in Washington DC. He is Professor of Zoology at Calcutta University and has been teaching Zoology for nearly thirty years at the postgraduate level.

Dr. Genaro Diarte-Plata is a professor and researcher with appointment at the candidate level at the Sistema Nacional de Investigadores (SNI) Consejo Nacional de Ciencia y Tecnología (CONACYT) Mexico. He is a biologist in Aquaculture at the Universidad Autónoma de Sinaloa, Mexico. He holds a master's in Natural Resources and Environment from the Instituto Politécnico Nacional (IPN CIIDIR-Sinaloa), Mexico. He has been a professor at the Instituto Politécnico Nacional, Mexico, since 2006. He is also a researcher for the Sinaloa System of Researchers and Technologists (SSIT-INAPI Sinaloa). He has a doctorate in Marine and Coastal Sciences from the Universidad Autónoma de Baja California Sur, Mexico. Dr. Diarte-Plata conducts investigations in mariculture, and culture and ecology of molluscs and crustaceans. As a professor-researcher within the IPN CIIDIR-Sinaloa he has participated in 48 research projects, given 11 seminars and 260 presentations at international and national conferences, and has written 5 book chapters and 19 indexed articles. He teaches Descriptive and Multivariate Statistics at the bachelor's level, and Aquaculture at the postgraduate level. He directs 15 undergraduate theses, 5 master's degrees, and 2 doctorates from various universities. He is also an arbitrator in the revision of manuscripts sent to international journals (ISI-JCR).

Dr. Ruth Escamilla-Montes holds a doctorate in Marine and Coastal Sciences with a specialty in Aquaculture. Belonging to Assistant Technologist of the Sinaloa System of Researchers and Technologists. Here areas of study include cultivation and ecology of molluscs and crustaceans. Over the last five years she has focused on the isolation and characterization of lactic acid bacteria, yeasts, and bacilli with probiotic potential and pathogenic bacteria, such as *Vibrio parahaemolyticus*, and their effects on the growth, survival, and expression of genes of the immune system and stress of marine organisms. She has participated in 11 research projects as well as presentations at national and international conferences. She has authored 4 book chapters and 28 articles in ISI Thomson–JCR international journals. She is also an arbitrator in the revision of manuscripts sent to international journals (ISI-JCR).

Contents

Preface

A complex process of biological evolution resulted in the birth of multiple species of plants and animals. The basis of a synthetic approach to studying evolution is integration of Mendelian genetics with the theory of natural selection. The concept of an adaptational paradigm rests on the premises of attributes of heritable adaptation and genetic functioning stabilized through the process of natural selection. Most of the extant forms of metazoa evolved during the Precambrian–Cambrian transition. Evolutionary biologists debated on issues like protist ancestry of primitive metazoa, the evolutionary process of multicellularity, and the nature of favorable environment, which facilitated the birth of primitive forms with multicellularity. However, evolution of bilateral anatomy, coelomic organization, and biological defense mechanisms are assumed to be the major milestones of metazoan evolution that led to speciation in many invertebrate groups. Invertebrates are thermo-conformers and uniquely adapted to multiple thermal conditions. Evolution of effective ectothermy enabled their cells and tissues to function in diverse thermal regimes within a homeostatic range. Invertebrates, which represent more than 90 percent of the fauna, exhibit a wide range of diversity and uniqueness in relation to their body plan, physiology, behavior, and preferences for habitat and food. They are suitably adapted and distributed in almost all ecological conditions of earth. Evolution and organisation of coelom is considered as an important trend of evolution culminating in the formation of multidimensional body forms and subsequent adaptation in different bioecological conditions. The resilience of physiological functioning on an ecological stage holds the key of evolutionary success of almost all of the positive variants in the pathway of adaptation and speciation.

In this book, authors highlight important ecophysiological aspects and management-related issues of invertebrates across the phylogeny. Researchers working in the frontier areas of ecophysiology and animal management present their findings and analyses from different points of view. Science of animal management, in recent years, has been gaining special attention, particularly under the backdrop of environmental contamination and climate change. Animal forms like sponge, corals, and molluscs have already been identified as the major victims of global warming, climate change, chemical contamination, and habitat loss-related crises. Undesirable and unpredictable shifts in ocean current, rain fall, and ecothermal characteristics are the current issues of concern to the scientists working in the fields of ecophysiology, distribution, aquaculture, and animal management.

This book is comprised of informative multidisciplinary articles under two sections entitled "Ecophysiology" and "Management." In the first chapter in the section on "Ecophysiology," authors report the key features of French shell farming technology where the matured garden snails are allowed to hibernate under controlled microenvironmental conditions. Issues of survival, dormancy, and hibernation of all economically important snails are addressed along with the phenomenon of Oblomovism, a less studied physiological feature of many molluscs. This particular chapter bears special significance in applied malacology. In another chapter, scientists report the ecological, biological, and genetic impacts of *Millepora* hydrocorals on coral reefs. Status of taxonomy, biogeography, ecology, and symbiosis of this less researched group of reef component are screened in depth. Strategies of reproduction, dispersal, and growth are examined in hydrocorals at the defined stage of ecological conditions. Authors also stress the genotypic and phenotypic analyses of coral to interpret the ecological and evolutionary basis of persistence of hydrocorals. Their role as ecological engineer is being challenged due to population decline. The next chapter in this section reports the chemoprotective and ameliorating potentials of an edible mollusc against cisplatin toxicity. Cisplatin and its metabolites exhibit genotoxicity by binding with DNA and thus appears to be physiologically hazardous to mammals. Pretreatment of experimental mice with molluscan extract indicate inhibition or reversal of the toxic effect of cisplatin. This novel report is significant in that it presents the biological importance of this gastropod in pharmacology and cancer therapy. The next chapter discusses schistosomiasis, which is a major parasitic disease of the human population of sub-Saharan Africa. The parasite is transmitted through freshwater snails, which serve as intermediate hosts of it. In this chapter, authors report the effects of the physicochemical profile of river water and the spatial distribution of the snail. The study was carried out around the delta of the Senegal River of Africa and the authors highlight the influence of selected water parameters including the concentrations of phosphates, nitrates, salinity, and conductivity on the relative density of the host snail. The study is significant for public health science, parasitology, and epidemiology.

The second section on “Management” represents the current status of technology and the scientific approach to sustainable management of selected invertebrate groups. Species groups with commercial and ecological significance are of general interest for environmental managers, aquaculturists, conservationists, and a section of biotechnologists. In the first chapter in this section, researchers report on the reproductive biology, seed production, and aquaculture strategy of the limpet. This chapter aims to develop a sustainable aquaculture technology for the Hawaiian limpet. Effective dietary manipulation is suggested for a better yield of this endemic species. Sponges of both freshwater and marine origin bear a significant role in ecology, ethnomedicine, and biotechnology. Many of the species are sources of biosilica and pharmacological compounds. Sponge fishery, in current years, has been gaining the attention of aquaculturists. An informative article on sponge fishery would enrich the understanding and knowledge base of the modern technologists of marine fishery. Application of cutting-edge technology for biological status evaluation of coastal invertebrates appears to be novel and unique. In one chapter, scientists highlight the efficacy and potential of field spectroscopy in screening the status of health of coral in situ. Hyperspectral signatures of live and dead corals were examined employing state-of-the-art technology. This report has significance in constructing a faunal map of coral and its subsequent conservation in the coastal ecosystem. In another chapter, scientists present a case study of formation and persistence of glass sponge reef. This type of reef is generally formed on glacial deposits. Use of barcoded stakes over a long period of time for identification and verification of location of the species is a novel feature of the study. Photographic documentation of growth, collapse, and regrowth were carried out in detail. The section of this book also contains a chapter stating the artisanal harvest of limpet and topshell at the north-eastern Atlantic. This important study reports the status of artisanal harvest operative in this region for a period of nearly three decades. This type of detailed characterization of gastropod harvesting helps in comparing future data under the backdrop of the increasing trend of resource utilization and species conservation. Assemblage of gastropod molluscs in the intertidal zone of Asry beach of the kingdom of Baharin was reported on the basis of field data gathered for a period of three years. Gastropods are the major intertidal fauna of this region. This intertidal ecosystem is characterized by sandy and rocky substrata. Rocky substratum of beaches experiences a low level of anthropogenic stress and human intervention and its biota appear to be tolerable to temperature change as indicated by a stable state of gastropod assemblage. Authors suggest routine monitoring of the biodiversity status for effective protection and conservation of the faunal assemblage along with their natural habitat.

This edited volume is a collection of chapters written on the basis of experimental and field-based scientific observations made by eminent scientists in their respective research fields. Recently, the disciplines of ecophysiology and management of invertebrates have been experiencing a shift in research paradigm. Availability and application of novel technology and sophisticated statistical tools led to the generation of newer sets of information in the field of invertebrate zoology. Biologists with diverse research backgrounds and interests have been participating in invertebrate research in a greater number than before. This unique compilation is expected to attract the new generation of students, researchers, and teachers towards the discipline of invertebrate biology.

Sajal Ray
University of Calcutta,
India

Genaro Diarte-Plata and Ruth Escamilla-Montes
Instituto Politecnico Nacional (CIIDIR Sinaloa),
Mexico

Section 1

Ecophysiology

Chapter 1

Indoor Hibernation of *Helix aspersa* Juveniles

George Andrei Draghici, Cristina Deheleana, Razvan Susan, Delia Berceanu-Văduva and Dragoş Nica

Abstract

The "Italian" outdoor snailfarming technology assumes that both mature and juvenile snails hibernate outdoor, protected by a thin sheet of unweaved coverlet (agryl sheet). In contrast, the "French" snailfarming technology implies that only mature brown garden snails (*Helix aspersa*) hibernate indoor, in strictly controlled microenvironmental parameters (temperature, humidity, and ventilation). This technology may also be viable for *H. aspersa* juveniles. Extremely high death rates occurring in Romanian outdoor snailfarms during colder winters (>80%) imposed the need to find alternative paths for a proper hibernation of *H. aspersa*. Using statistical analyses, close surveillance of technological flow, and controlled micro-environmental parameters, we assessed the possibility to adapt indoor hibernation for *H. aspersa* juveniles. The experiments lasted for 2 years (2006–2008) and were carried out on 34,000 *H. aspersa* juveniles and 15,000 mature ones, using different technological flows and microenvironmental parameters (temperature, humidity, and ventilation). They were performed in two stages and involved five case studies, conducted independently in three different locations: Floreşti (Mehedinţi county), Sântuhalm (Hunedoara county), and Muntenii de Sus (Vaslui county). The first stage tested the hypothesis in relation to survival rate of mature snails, *H. aspersa*, in the same conditions, whereas the second stage improved the technological flow, before its extensive application. We demonstrated that noncontrolled microclimate parameters (temperature, humidity, and ventilation) and the use of straw as hibernation support induced significant differences ($P < 0.01$) concerning death levels of *H. aspersa* juveniles as compared to their indoor hibernation in semicontrolled microclimate (temperature and ventilation). In the same hibernation microclimate, mature snails exhibited higher survival levels than the juvenile ones, irrespective of technological flow and origin ($P < 0.0001$). We also demonstrated that juveniles' weight loss displays a relatively constant variation (16.33–20.51%). In addition, the correlations between the individual average weight before and after hibernation were described by the same logarithmic regression. Furthermore, significantly higher survival rates of *H. aspersa* juveniles ($P < 0.0001$) have been registered when they had not been awakened during hibernation. Finally, we proved that indoor hibernation of *H. aspersa* juveniles in strictly controlled microenvironmental parameters (temperature, humidity, and ventilation) could represent a viable technology that improves the technological flow in outdoor snailfarming during wintertime in colder climates.

Keywords: microenvironment, snailfarming, hibernation, technology, monitored

1. Introduction

In a continental climate, characterized by higher rainfall levels than on countries with tradition in snailfarming (France, Italy, Spain, Greece), in Romania this activity registered a booming development during the 2003–2007 time period [1]. Thus, in 2006, according to the International Institute of Snail Farming from Cherasco (Italy), Romania ranked second in the world concerning the number of outdoor snailfarms (>1000) and their sown area. The "French" snailfarming technology implies that the snails are bred in captivity, and juveniles are introduced early in the spring in outside fattening pens, wherein they are fed primarily a combination of concentrated fodders [2]. As a result, most snails reach adulthood from 6 to 8 months, and in autumn they are sold as final product. Only a small proportion of adult gastropods is kept as reproductive herd for the next year productive cycle and hibernate in strictly controlled indoor environment [3]. The immature juveniles are not gathered; therefore, they are let to survive outside during wintertime, without any additional protection [4]. In contrast, the "Italian" snailfarming technology snails employs the biological cycle of raising and growing snails in open pastures of fresh vegetables [5]. A typical farm is organized in pens with precise destinations: 60% for breeding and 40% for fattening [6]. The fattening pens are used starting from the second year of activity onward, when after hibernation, snails are transferred from the breeding pens into the fattening pens [7]. When winter arrives, snail of many sizes, starting from hatchlings to adult ones, are found inside the pens [1]. The solution used for snail hibernation relies on trimming the vegetation inside the pens to 20 cm in height, whereas the pens are covered with unweaved coverlet (weight = 18–25 grams per square meters, i.e., g/m^2)—material also known as agryl sheet [8, 9].

High death rates have occurred in snailfarms all around Romania during the winter of the year 2006, proving that the standard outdoor hibernation technology is not well suited for colder climates (temperate continental climate). As a result, our research focused on finding some alternative paths for a proper hibernation of *H. aspersa* in colder climates. One solution was the development of "sandwich" system—a protective structure based on the nonconducting properties of the straw, on soil thermic inertia, and the insulator properties of nylon sheet. This system was tested at micropilot level in 2006 [9] and was extensively used in outdoor snailfarms [1]. However, two additional possibilities were also tested: indoor hibernation of *H. aspersa* juveniles and indoor rearing of *H. aspersa* juveniles during wintertime. It is known that during hibernation, the gastropods' vital functions decrease to subsistence level [10, 11] and the shell aperture is sealed with one or several epiphragms [12], allowing these terrestrial mollusks to survive in a stage of dormancy up to 4–6 months [13]. Indoor hibernation of mature snails, *H. aspersa*, in a controlled environment, temperature 2–6°C and humidity 70–80% [14], allows at least 80% of them successfully to pass overwinter [4]. We considered that indoor hibernation could represent a possible solution for *H. aspersa* juveniles, if this approach can be adapted for their physiological needs. The optimal survival level (Slo) of mature snails during hibernation (80%) was considered as a benchmark to assess the viability of this technology for *H. aspersa* juveniles. The experiments of this pilot exploratory study lasted 2 years and were performed in two distinct stages, in three snailfarms, and on 34,000 juvenile specimens of *H. aspersa* and 15,000 mature specimens of *H. aspersa*, using different technological flows and microenvironment parameters (temperature, humidity, and ventilation). Within the first stage, we conducted mixed experiments, using both mature snails and juvenile snails to evaluate the influence of technological flow and variable microclimate parameters on the snails' intermediate survival rate (Sli) and final survival rate (Slf).

Then, we analyzed weight variation of *H. aspersa* juveniles during indoor hibernation (Wl, Wl%). The second stage assessed the viability of this novel approach before its extensive application, in relation to Slo and death levels of *H. aspersa* juveniles recorded during the first stage of our pilot exploratory study.

2. Materials and methods

The experiments of this pilot exploratory study were conducted in three snailfarms chosen based on their location, technological flow, and microenvironment parameters: Floreşti (Mehedinţi county; latitude, 44°75'; longitude, 22°92'), Sântuhalm (Hunedoara county; latitude, 45°85'; longitude, 22°96'), and Muntenii de Sus (Vaslui county; latitude, 46°70'; longitude, 27°76'). The farms were carefully monitored since their implementation: 2005 (Muntenii de Sus) and 2006 (Floreşti, Sântuhalm). The reproductive herd was imported from Italy. The data were carefully monitored and recorded into technological evidence files. Next, they were used for five case studies (**Table 1**) depending on location and snail size: Cs1 (Floreşti, juvenile *H. aspersa* juveniles); Cs2 (Floreşti, mature *H. aspersa* snails); Cs3 (Sântuhalm, juvenile *H. aspersa* snails); Cs4 (Sântuhalm, mature *H. aspersa* snails); and Cs5 (Muntenii de Sus, juvenile *H. aspersa snails*). Field observations performed in 40 outdoor snailfarms from 2004 to 2007 indicated that, under Romanian pedoclimatic conditions, the mating season began in June, with most juveniles hatching in September. This does not allow the juveniles to exceed 1.0 cm in shell diameter till hibernation, and therefore, we considered that the snail size was homogeneous enough to provide accurate data. Moreover, before being hibernated, they were carefully selected by using a fine strainer (φ mesh = 1.15 cm). Thus, the term "juveniles" defines in this study young snails with shell diameter up to 1 cm. The indoor hibernation experiments monitored the survival levels of juvenile and mature specimens of *H. aspersa* in relation to three primary parameters: temperature, humidity, and ventilation (**Table 1**).

2.1 Hypothesis testing

First, two distinct locations were selected for these studies: Floreşti and Sântuhalm. Two lots were sampled from each location, one containing only juvenile *H. aspersa* snails and another only mature *H. aspersa* snails: Floreşti (Cs1, Cs2) and Sântuhalm (Cs3, Cs4). About 5 kg of juvenile *H. aspersa* snails were collected for each location. Five lots, about 100 g each (S_1–S_5), were aleatory collected for each location. Next, the number of juveniles was counted for each lot. Then, we estimated the individual average weight before hibernation (Wb) for each sample as the ratio between the total weight and the number of juveniles. After that, we estimated the number of hibernating juveniles (Nb) as the ratio between the total sample weight and Wb. During indoor hibernation, high death rates were recorded for juveniles, with each sample taken into account weighing after hibernation about 50 g. The individual average weight (Wa) and the number of juveniles for each sample (Na) after hibernation were assessed in the same manner as for before hibernation. Then, the survival level was calculated as the ratio (%) between Nb and Na. We also determined the weight loss during hibernation (Wl) for each sample as the difference between Wb and Wa. Then, the percentage weight loss (Wl%) was calculated as the ratio between Wl and Wa. For their mature

<table>
<tr><th></th><th>Sampling data Number/(weight)</th><th>Intermediary control dead snails</th><th>Awakening (live snails)</th><th>T (°C)</th><th>U%</th><th>Ventilation</th></tr>
<tr><td>Cs1</td><td>November 17, 2006
5025 g
(≈10,255 pcs.)</td><td>January 12, 2006</td><td>March 13, 2007
3670 juveniles</td><td>Constant 2–5°C</td><td>Variable 60–75%</td><td>Partially controlled</td></tr>
<tr><td></td><td colspan="6">Hibernation: Juveniles were introduced in a purging case (length, 100 cm; width, 100 cm; height, 25 cm), built from galvanized wire net (φ mesh = 0.3 cm), placed on a metallic frame, in a 5-cm-thick layer</td></tr>
<tr><td></td><td colspan="6">Intermediary control: Snails were awaken from hibernation and fed with concentrated fodder and minced carrots</td></tr>
<tr><td></td><td colspan="6">Microclimate: Undercroft, thermically insulated with extruded polystyrene, 5 cm thick
• Temperature and humidity were constantly monitored by using two electronic thermohigrometers, no light</td></tr>
<tr><td></td><td colspan="6">• Starting from January 12, 2007, a Ufesa VP3801 ventilator was installed. This ventilator, capable of gyratory movement to an angle of 80°, was used 5 days/weak and 2 hours/day</td></tr>
<tr><td>Cs2</td><td>November 17, 2006
6500 pcs.</td><td>January 12, 2007
640 pcs.</td><td>March 13, 2007
5906 pcs.</td><td>Idem Cs1</td><td></td><td></td></tr>
<tr><td></td><td colspan="6">Hibernation: Snails were introduced in a purging case (length, 80 cm; width, 80 cm; height, 30 cm) was built of galvanized wire net (φ mesh = 1.0 cm), placed on a metallic frame, in a 20-cm-thick layer</td></tr>
<tr><td>Cs3</td><td>November 11, 2006
5050 g
(≈12,949 pcs.)</td><td>January 1, 2007</td><td>March 7, 2007
724 pcs</td><td>Variable
−1 → + 8°C</td><td>Variable 55–90%</td><td>Variable</td></tr>
<tr><td></td><td colspan="6">Hibernation: Hibernation in a cage (length = 100 cm, width = 100 cm, height = 25 cm), built of glass fiber net (φ mesh = 0.1 cm) placed on a wooden frame, previously disinfected with a 10% quick lime solution. Then, they were uniformly placed in layer, 3 cm in thickness, between two layers of straw, the upper one 3 cm thick and the lower one 5 cm thick</td></tr>
<tr><td></td><td colspan="6">Intermediary control: January 1, 2007
Microclimate: Non-insulated outdoor storage, built of burnt bricks</td></tr>
<tr><td></td><td colspan="6">• Temperature and humidity (50–85%) were constantly monitored by using two electronic thermohigrometers</td></tr>
<tr><td>Cs4</td><td>November 11, 2006
8500 pcs.</td><td>January 1, 2007
2000 pcs.</td><td>March 7, 2007
5800 pcs.</td><td>Idem Cs3</td><td></td><td></td></tr>
<tr><td></td><td colspan="6">Hibernation: Snails were introduced in a purging cage (length, 200 cm; width, 150 cm; height, 50 cm) built of glass fiber net (φ mesh = 1.0 cm), placed on an oak wood frame. Then, they were uniformly placed between two layers of straw, the upper one 15 cm thick and the lower one 10 cm thick</td></tr>
<tr><td>Cs5</td><td>November 11, 2007
5, 050 g (≈10,100 pcs.)</td><td>—</td><td>March 18, 2008
6837 juveniles</td><td>Constant 2–5°C</td><td>Constant 70–75%</td><td>Controlled</td></tr>
<tr><td></td><td colspan="6">Hibernation: The juveniles were put in a multileveled structure, made up of three wooden boxes (length, 80 cm; width, 40 cm; height, 25 cm). To build the hibernation structure, the boxes were turn upside down, put one over another, and then covered on the side and lower part with a fine mosquito net
Microclimate: Undercroft, thermically insulated with extruded polystyrene, 5 cm thick</td></tr>
<tr><td></td><td colspan="6">• Temperature was constantly monitored by using two electronic thermometers. Air relative humidity was controlled with a humidifier Venta Airwasher LW 14 from 70 to 75%. A Ufesa VP3801 ventilator was used 7 days/week and 1 hour/day, to provide a proper aeration</td></tr>
</table>

Table 1.
Indoor hibernation technological flow for the five case studies (Cs1–Cs5).

counterparts, only the intermediate and final survival levels were registered (Slf, Sli) as the percentage ratio between Na and Nb. The farms were monitored from September 2006 to April 2007, and all the data were carefully monitored and recorded in the technological evidence files (**Table 1**).

2.2 Technology optimization

The first stage data were used to optimize this technology in a study performed from October 2007 to March 2008 (**Table 1**) in a snailfarm located in Muntenii de Sus (Cs5). All the procedures were identical with those used in the study cases Cs1 and Cs3 (**Table 1**). The only exception was that the post-hibernal samples weighed about 75 g and not 50 g, like in the previous cases.

2.3 Statistical analysis

The hibernation efficiency (**Table 2**) was assessed based on the snail survival rate. To estimate the potential influence of origin and technological flow on juveniles' weight loss during wintertime (Wl), we analyzed all the quantitative indicators (individual weight, weight loss, snail number/known weight) by descriptive (**Figure 1**) and nonparametric statistical tests. First, we assessed the distribution normality (Anderson-Darling test) for Wb, Wl, Wa, Na, and Nb for all the samples (df = 1, n = 5/sample). After that, using a Kruskal-Wallis test with error Bonferroni correction (two-tailed, df = 2, n = 5/sample), we estimated Wb, Wa, Na, Nb, and Wl variations for juvenile snails (Cs1, Cs3, Cs5). Then, correlation analysis was performed to find whether these relationships between Wb and Wa displayed strong correlations among themselves. After that, we aimed to find the most appropriate function able to describe accurately these relationships. Thus, several attempts were conducted by using nonlinear regression. Finally, we chose the formulas that provided the highest precision (R^2) for all the samples taken into account.

	Cs1	Cs2	Cs3	Cs4	Cs5
			Before hibernation		
Wb	0.49 ± 0.06	—	0.39 ± 0.06	—	0.50 ± 0.04
Nb	205.00 ± 25.77	—	262.20 ± 40.61	—	200.00 ± 14.98
			After hibernation		
Wa	0.41 ± 0.04	—	0.31 ± 0.05	—	0.41 ± 0.03
Na	123.40 ± 14.77	—	160.80 ± 21.47	—	175.6 ± 24.82
Wl	0.087 ± 0.027		0.075 ± 0.023		0.089 ± 0.009
Wl%	16.33%		20.51%		18.00%
			Intermediary control		
Sli	—	91.62%	—	76.46%	—
Dri	—	8.38%	—	23.53%	—
Final control					
Slf	35.78%	78.55%	18.06%	68.34%	67.69%
Drf	64.22%	21.45%	81.94%	31.66%	32.31

Table 2.
Descriptive statistics (mean individual weight, number/known amount, and weight loss during winter time; X ± SE, n = 5), weight loss (%), and survival parameters for the five case studies (Cs1–Cs5).

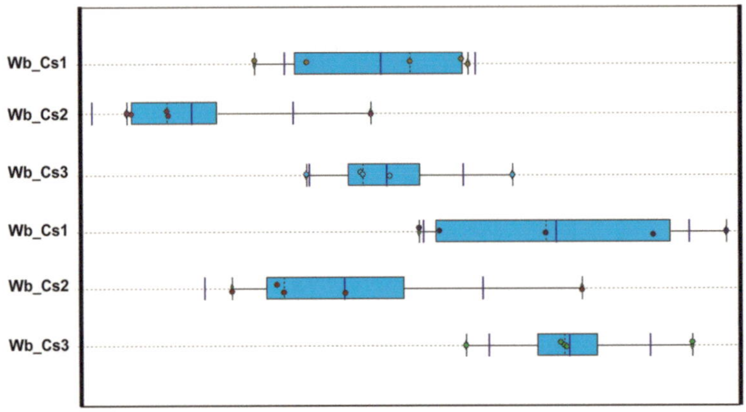

Figure 1.
Individual average weight (WA) variability.

Death rates were analyzed by using a χ^2 test (df = 1, two-tailed). To reduce the error in approximation, we adjusted χ^2 according to Yates' correction for continuity [15]. First, in 2007 we assessed the cumulated actions of size and technological flow on survival rate of *H. aspersa* juveniles during indoor hibernation in relation to mature snail mortalities (Cs1 vs. Cs2, Cs3 vs. Cs4). Next, we estimated the effect of technological flow on juveniles survival levels (Cs1 vs. Cs3). Then, we analyzed the impact of technological flow and microenvironment parameters on Slf and Sli for mature *H. aspersa* snails (Cs2, Cs4). After that, all the data were analyzed in relation to the optimal survival level for the indoor hibernation of mature snails, *H. aspersa* (80%). The results were used to adapt the indoor hibernation technological flow and to optimize it for the juvenile snails. These principles were put into practice for Cs5. Finally, in 2008 the survival level recorded in Cs5 was assessed in comparison to the juveniles' indoor hibernation attempts from 2007 (Cs1, Cs3). Before making a final conclusion, the Cs5 results were finally compared with Slo. Since this was a pilot exploratory study, no sample size calculation was needed.

3. Results

3.1 Hypothesis testing (Cs1–Cs4)

The Anderson-Darling test proved an abnormal distribution ($P > 0.05$) for all the quantitative parameters taken into account: mean individual weight, weight loss during hibernation, and snail number/known weight. Descriptive statistics (**Figure 1**, **Table 2**) revealed that before hibernation, Wb variability was almost equal for Cs1 and Cs3. In contrast, after hibernation Wa variability was higher for Cs3 (**Figure 1**, **Table 2**). Strong correlations between Wb and Wa are found for both Cs1 ($P = 0.013$; $R = 0.90$) and Cs3 ($P = 0.005$; $R = 0.95$). Irrespective of location, the Kruskal-Wallis test exhibited no significant influences ($P > 0.05$) of origin and indoor hibernation technological flow on Wb and Nb variation. Similar data were found for Wa, Na, and Wl variation ($P > 0.05$). In addition, the highest precision for correlations between Wa and Wb for both Cs1 and Cs3 was achieved by using a logarithmic regression described by the same base formula:

$$Wa = \mathrm{a}/\{1 + \log\left[-\mathrm{b} - (\mathrm{c} * Wb]\right\} \quad (1)$$

DOI: http://dx.doi.org/10.5772/intechopen.88732

where *Wa* = mean weight after hibernation (g), *Wb* = mean weight before hibernation (g), and *a*, *b*, *c* = constant values (**Figure 2A**, **B**). The constants and the nonlinear correlation coefficients displayed the following values: Cs1 (R = 0.99; $a \approx 0.46$; $b \approx 7.21$; $c \approx 19.66$) and Cs3 (R = 0.82; $a \approx 0.46$; $b \approx 3.02$; $c \approx 2.49$).

The data recorded in the technological files revealed that, for Cs3 and Cs4, the problems started from January 1, 2007, when suddenly the outdoor temperature increased over 5°C and abundant rainfall (slushes) were recorded. Because the storage had no thermic insulation, the air humidity exceeded 85%, water condensated on the storage walls, the straw soaked, and the snails, especially the juvenile ones, started to awaken from hibernation. As a result, the straw were removed, and the dead snails were also drawn away. This action limited the death rate, but at the same time, it induced the restart of their metabolic cycle, especially for juvenile snails. This behavior of *H. aspersa* juveniles was attributed to their partial oblomovism. Under the same hibernation conditions, the size displayed a significant influence on Slf: Cs1 vs. Cs2 (χ^2 = 35.63, $P < 0.0001$) and Cs3 vs. Cs4 (χ^2 = 49.49, $P < 0.0001$). Different technological flows and microenvironment parameters (temperature, humidity, and ventilation) significantly influenced Slf for juvenile snails in Cs1 and Cs3 (χ^2 = 7.11, $P < 0.01$). Significant differences were also recorded regarding mature snails Slf (χ^2 = 7.47, $P < 0.01$) and Sli (χ^2 = 7.37, $P < 0.01$) in Cs2 and Cs4. Possible reasons were attributed to the cumulated effect of straw moistening, variable temperatures, and increased humidity. As a result, we eliminated from the technological flow the use of straw as hibernation support and considered unrecommended the fluctuation of microclimate parameters (temperature, humidity, and ventilation) inside the hibernation chamber. In addition, there seemed to be a positive relationship between snail size and tolerance to indoor hibernation.

At that time, no data were available in literature or in practice concerning the maximum period that allows juveniles to successfully survive during wintertime. Thus, Cs1 snails were awakened form hibernation on January 12, 2007. Next, they were fed with concentrated fodder and minced carrots, and after that, they were reintroduced to hibernation. After hibernation, the comparative statistical analyses revealed significant differences between Slf and Slo for the juvenile snails, in both locations: Cs1 (χ^2 = 38.31, $P < 0.0001$) and Cs3 (χ^2 = 74.23, $P < 0.0001$). Thus, it become obvious that during wintertime the juveniles must hibernate continuously, whereas their awakening and feeding must also be excluded from the technological flow. In addition, the higher survival rate and the lower Wb variability registered in Cs1 supported the idea that the technological used in this case represents the base for the future trials. In contrast, there were no significant differences in Slo between Cs2 (χ^2 = 0.01, $P > 0.05$) and Cs4 (χ^2 = 2.97, $P > 0.05$). These data confirmed mature snails' capacity to survive better during indoor hibernation than their younger counterparts. Although the Slf recorded for Cs2 and especially for Cs4 were lower

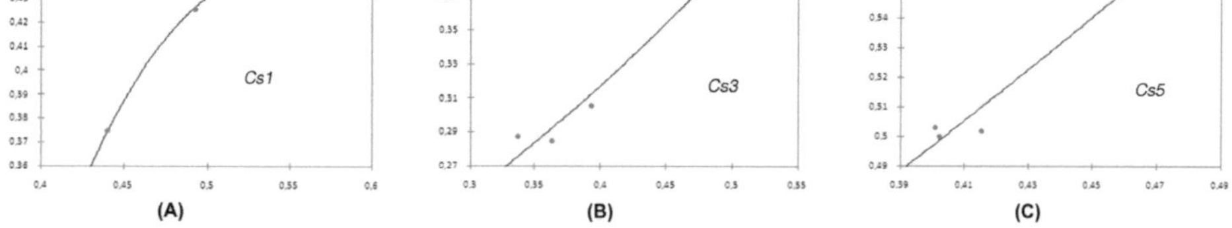

Figure 2.
Constant values (A, B, C).

than Slo, they were considered acceptable for indoor hibernation of mature snails, *H. aspersa*, considering that the microenvironment parameters were not totally identical with those used in the standard technology.

3.2 Technology optimization (Cs5)

The same abnormal distribution ($P > 0.05$) for all the quantitative parameters (mean individual weight, weight loss during hibernation, and snail number/known weight) was revealed for Cs5 (**Figure 1**, **Table 2**). Descriptive statistics (**Figure 1**, **Table 2**) revealed that Wa and Wb presented a lower variability for Cs5 than both Cs1 and Cs3. Correlation analysis showed that, similar to Cs1 and Cs3, there were strong correlations between Wb and Wa for Cs5 ($P = 0.004$; $R = 0.94$). Moreover, the Kruskal-Wallis test displayed no significant influences on Wb ($P > 0.05$) and Nb variation ($P > 0.05$). Similar results were also found for the nonparametric multiple analyses of Wa, Na, and Wl variation ($P > 0.05$). Moreover, the highest precision in estimating the correlation between Wa and Wb was achieved by using a logarithmic regression described by the same base formula as in Cs1 and Cs3 (**Figure 2C**). The constants and the nonlinear correlation coefficient displayed the following values: $R = 0.96$, $a \approx 1.33$, $b \approx 7.21$, and $c \approx 2.73$.

Concerning Slf, Cs5 proved significant differences in comparison to both Cs1 ($\chi^2 = 19.31$, $P < 0.0001$) and Cs3 ($\chi^2 = 48.02$, $P < 0.0001$). In contrast, no significant differences were found when Slf comparative analyses were performed in relation to Cs2 ($\chi^2 = 2.46$, $P > 0.05$) and Cs4 ($\chi^2 = 0.01$, $P > 0.05$). In addition, similar data were found in relation to Slo ($\chi^2 = 3.31$, $P > 0.05$). Moreover, weight loss overwinter in totally controlled condition (Cs5) presented a lower variability than for Cs1 and Cs3 (**Figure 1**). As a result, the technological flow followed in Cs5 was considered proper for indoor hibernation of *H. aspersa* juveniles because it provided the highest survival rate and the lowest weight variation during wintertime.

This study demonstrated without doubt that indoor hibernation of *H. aspersa* juveniles is a viable technology, which could be successfully used to improve the technological flow of outdoor snailfarming in colder climates. A successful implementation implies that microenvironmental factors (temperature, humidity, ventilation) are constant, whereas during hibernation the juveniles are not fed. However, mature snails seem to present a better tolerance to indoor hibernation; therefore this process must be more carefully controlled for the juvenile *H. aspersa* snails than for their mature counterparts.

4. Discussion

The key factors triggering land snail dormancy are temperature decrease [16], photoperiod diminishing [17], and low humidity [18]. For *H. aspersa*, extensive research proved that this process is controlled mainly by photoperiod [19], whereas temperature may determine its duration [20]. Although our experiments were conducted in the absence of light, regardless of size, the snails successfully hibernated for 100–110 days. However, we suggest that the cumulated actions of environmental factors (photoperiod, temperature, humidity) will easily induce snail hibernation than their simulation in a controlled microenvironment. As a result, to provide a successful indoor hibernation, the collection moment must be carefully chosen. Based on previous studies concerning *H. aspersa* prehibernal behavior in Romania pedoclimatic conditions [9], we considered that the proper moment to pick up the snails within the farm is early in the fall (September-October) when there is a low day-night amplitude, the day length shortens below 10 hours, the temperature slowly and constantly decreases, and the soil humidity is moderate (<70%).

Oblomovism is well known in the world of mollusks [21]. This term came from the homonym novel written by Ivan Goncharov and is used to describe someone who exhibits the personality traits of sloth. Thus, during their life, snails pass through short periods of great activity, essential for building up their reserves, which alternates with frequent periods of inactivity, when they are sleeping or pending the favorable weather. Taylor [22] considered juvenile snails less sensitive to cold and thus less inclined to hibernation; therefore it was considered that they exhibit a partial oblomovism. However, recent studies proved that for *H. aspersa* freezing tolerance abilities vary converse with size [23]. This tendency was also noticed in our previous studies [9]. We suggested that this behavior has another more plausible explanation than the one proposed previously. Thus, the biological clock of juveniles *H. aspersa* is delayed as compared to their mature counterparts so they will start to hibernate later and will exhibit a long active cycle. Our studies also revealed that when temperature increased over the critical point of activity, +5°C for *H. aspersa* [24], this rule was also valid for indoor hibernation. The length of hibernation affects temperature-induced spermatogenic multiplication in *Helix aspersa* [25]. Additionally, a proper hibernation increases the reproductive activity and fecundity of *H. aspersa* [26]. As a result, a properly timed hibernation is a key factor in an outdoor snailfarm rentable gestion.

Our findings proved that juveniles displayed, regardless of the technology flow and origin, a relatively constant variation of weight after 100–110 days of indoor hibernation. Although in the wild, snails displayed variable losses of weight in relation to climatic factors [1], we considered that snail adaptation to hibernation throughout their long evolution and the controlled microenvironment allowed them to pass overwinter with a relatively constant weight loss during wintertime. Thus, we consider that this technology might be used in outdoor snailfarms located in colder areas with temperate continental climates as efficient alternative to the simple outdoor hibernation.

Author details

George Andrei Draghici[1]*, Cristina Deheleana[1], Razvan Susan[2], Delia Berceanu-Văduva[2] and Dragoş Nica[1]

1 Faculty of Pharmacy, "Victor Babes" University of Medicine and Pharmacy Timisoara, Timisoara, Romania

2 Faculty of Medicine, "Victor Babes" University of Medicine and Pharmacy Timisoara, Timisoara, Romania

*Address all correspondence to: draghici.george-andrei@umft.ro

References

[1] Nica D. Researches regarding the morphophysiology of the gastropods and the farming of the Little grey snail (*Helix aspersa* Muller) in the pedoclimatic conditions of our country [thesis]. Timisoara, Romania: USAMVBT; 2009

[2] Bonnet JC, Aupinel P, Vrillon JL. L'escargot *Helix aspersa*: Biologie. Institut National de la Recherche Agronomique: Elevage; 1990

[3] Aubert C, Simoncelli C. Memento de l'éleveur d'escargots. Institut Technique D'aviculture; 2003

[4] Mathieu Y. Créer son élevage d'escargots. Paris: éditions Rustica; 2005

[5] Elmslie LJ. Snails and snail farming. World Animal Review. 1982;**41**:20-26

[6] Elmslie LJ. The growth of snail in breeding [Crescita della chiocciola in allevamento]; 1992

[7] Elmslie LJ. The potential for snail farming. 1983;**25**(1):35-37. In Progress in Agriculture Foston. Stevens JGR, editors

[8] Griglione N. Conservatione invernale di *Helix pomatia*. Elicicoltura. 1994;**4**: 4-5

[9] Avagnina G. Elicicoltura. Cherasco: Istituto Internacional de Helicicultura; 2003

[10] Nica D, Bura M, Munteau XC, Pătruică S. Study regarding the Brown garden snail (*Helix aspersa*) wintertime protection in open air rearing technology. USAMVB Timisoara Scientifical Papers Animal Sciences and Biotechnologies. 2007;**40**(2):638-644

[11] Ansart A, Vernon P, Daguzan J. Elements of cold hardiness in a littoral population of the land snail *Helix aspersa* (*Gastropoda*: *Pulmonata*). Journal of Comparative Physiology B. 2002;**172**: 619-625. DOI: 10.1007/s00360-002-0290-z

[12] Vorhaben JE, Klotz AV, Campbell JW. Activity and oxidative metabolism of the land snail *Helix aspersa*. Physiological Zoology. 1984; **57**(3):357-365

[13] Daguzan J. Contribution à l'élevage de l'escargot Petit-Gris: *Helix aspersa* Müller (Mollusque Gastéropode Pulmoné stylommatophore). III-Elevage mixte (reproduction en bâtiment contrôlé et engraissement en parc extérieur) activité des individus et évolution de la population juvénile selon la charge biotique du parc. Annales de Zootechnie. 1985;**34**:127-148

[14] Daguzan J. Contribution à l'élevage de l'escargot Petit-Gris: *Helix aspersa* Müller (Mollusque Gastéropode Pulmoné stylommatophore). II-Evolution de la population juvénile de l'éclosion à l'âge de 12 semaines, en bâtiment et en conditions d'élevage contrôlées. Annales de Zootechnie. 1982;**31**:87-110

[15] Sheskin DJ. Handbook of Parametric and Nonparametric Statistical Procedures. Taylor & Francis, Inc.; 2007. DOI: 10.1201/9781420036268

[16] Aupinel P, Daguzan J. Etude du rôle de la photopériode sur l'activité métabolique des jeunes escargots «Petit-gris» (*Helix aspersa* Müller) et mise en évidence d'une phase photosensible. Haliotis. 1987;**19**:47-55

[17] Bailey SER. Circannual and circadian rhythms in the snail *Helix aspersa* Müller and the photoperiodic control of annual activity and reproduction. Journal of Comparative Physiology. 1981;**142**:89-94. DOI: 10.1007/BF00605480

[18] Lazaridou-Dimitriadou M, Saunders DS. The influence of humidity, photoperiod, and temperature on the dormancy and activity of *Helix lucorum L*. (*Gastropoda*, *Pulmonata*). Journal of Molluscan Studies. 1986;**52**:180-189. DOI: 10.1093/mollus/52.3.180

[19] Ansart A, Vernon P, Daguzan J. Photoperiod is the main cue that triggers supercooling ability in the land snail, *Helix aspersa* (*Gastropoda*: *Helicidae*). Cryobiology. 2001;**42**:266-273. DOI: 10.1006/cryo.2001.2332

[20] Bailey SER. The photoperiodic control of hibernation and reproduction in the land snail *Helix aspersa* Müller. Journal of Molluscan Studies. 1983;**12**: 2-5. DOI: 10.1093/oxfordjournals.mollus.a065744

[21] Boss KJ. Oblomovism in the *Mollusca*. Transactions of the American Microscopical Society. 1974;**93**(4): 460-481. DOI: 10.2307/3225152

[22] Taylor JW. Monograph of the Land and Freshwater *Mollusca* of the British Isles. 3. Zonitidae, Endodontidae, Helicidae. Leeds: Taylor Brothers; 1906–1914

[23] Ansart A, Vernon P. Cold hardiness abilities vary with the size of the land snail *Cornu aspersum*. Comparative Biochemistry Physiology A. 2004;**139**: 205-211. DOI: 10.1016/j.cbpb.2004.09.003

[24] Capinera JL. Handbook of Vegetable Pests. San Diego: Academic Press; 2001

[25] Jess S, Marks RJ. Effect of temperature and photoperiod on growth and reproduction of *Helix aspersa* var. maxima. The Journal of Agricultural Science. 1998;**130**(3):367-372. DOI: 10.1017/S0021859698005346

[26] Bonnefoy-Claudet R, Deray A. Influence de Ia duree d'hibemation sur l'activite reproductrice de l'Escargot *Helix aspersa* Müller. Comptes Rendus des Seances de la Societe de Biologie et de Ses Filiales. 1984;**178**:442-449

Chapter 2

Ecology, Biology and Genetics of *Millepora* Hydrocorals on Coral Reefs

Caroline E. Dubé, Chloé A.F. Bourmaud, Alexandre Mercière, Serge Planes and Emilie Boissin

Abstract

Coral reefs are one of the most productive and diverse ecosystems on Earth. However, climate warming is occurring at an unprecedented rate and has negatively affected coral reefs worldwide. Evaluating the life history of reef-building species carries important implications for coral reef conservation. This chapter examines the taxonomy, biogeography, ecology, symbiosis, morphology and reproductive biology of *Millepora* hydrocorals, an important but relatively unstudied component of coral reefs. An emphasis is also placed on the influence of variable reef environments on *Millepora* life history traits, providing a fascinating opportunity to study the interplay between ecology and evolution. Special attention is given to ecological and evolutionary benefits of asexual versus sexual reproduction in the maintenance of genetic and phenotypic diversity. Lastly, this chapter discusses whether life-history strategies of *Millepora* hydrocorals and tolerance to different stressors can influence their ability to adapt and survive to future climate change, and other natural and anthropogenic disturbances.

Keywords: *Millepora*, coral reefs, taxonomy, biogeography, symbiosis, morphology, reproduction, population genetics

1. Introduction

1.1 Coral reefs: biodiversity and threats

Coral reefs were formed only 230 million years ago and are largely limited to warm shallow waters [1], yet they are among the most biologically diverse and economically important marine ecosystems. Coral reefs do not only shelter thousands of species; they also provide critical services to humans, including fisheries, coastal protection, medicines and tourism activities [2–4]. The economic value of coral reefs worldwide has been estimated to be around 30 billion US$ of net benefit per year [5]. Often called the rainforest of the sea due to their outstanding biodiversity, coral reefs only cover less than 0.1% of the ocean seafloor [6, 7] or approximately 5% that of rainforest areas [8]. Coral reefs thrive under nutrient-poor and oligotrophic waters [9–11], but yet harbour more than 25% of all known marine species [12, 13]. This ecosystem is sustained through efficient nutrient recycling strategies developed by corals [14] and algae [15], the primary reef producers, and other key organisms, i.e., microbes [16] and sponges [17]. In coral reef ecosystems, many calcifying benthic organisms contribute to reef accretion and build the complex and massive three-dimensional structure of reefs, including scleractinian corals, the major habitat architects, and hydrocorals [18]. These reef-builders are key components of coral reef health and biodiversity as they offer food, shelter and nurseries for thousands of reef-dwelling organisms and fishes [19].

Reefs are dynamic systems that are frequently punctuated by perturbations [20]. For instance, human activities can alter both global (climate change associated with CO_2 emissions) and local reef health (e.g. coastal habitat loss, pollution, sewage, overfishing and invasive species) [21, 22]. As a consequence, several reports of coral reef declines have been recorded, averaging 30–50% reductions in reef cover globally [4, 23, 24], including recent losses of coral cover following the multiple global bleaching events that occurred between 2014 and 2017 [25–27]. Since coral reefs are integrated ecosystems, declines in reef-building corals are often accompanied with declines of other species, such as many coral reef fishes [28], further hampering their capacity to deliver important ecosystem services to more than 500 million people [23, 29]. Yet, reef corals and the ecosystem they create can recover and a key factor underpinning such recovery is the ability of coral species to grow back, to maintain or renew their populations. Such an ability to respond to acute and chronic stressors in coral species is often linked to morphological traits, reproductive strategies and symbiont partners (among others). As we progress further into the Anthropocene, understanding and predicting these stress responses require prior knowledge on the life history traits of keystone reef corals, and some assessment of the influence that environmental changes may have on those traits.

1.2 *Millepora* hydrocorals

To date, the vast majority of studies on species' life history traits in coral reefs have mainly focused on scleractinian corals due to their key role in providing much of the habitat framework and structural complexity of reefs [30–34]. The extent to which other non-scleractinian reef-building organisms might rescue reef populations in response to environmental change is largely unknown. More information on such organisms is therefore needed. *Millepora* hydrocorals, known as fire corals because of their painful sting via toxic nematocysts [35, 36], are important components of reef communities where they, similar to scleractinian corals, contribute to reef accretion and community dynamics [37, 38]. *Millepora* species are Hydrozoans (Medusozoans), and together with hermatypic corals (Anthozoans), belong to the phylum Cnidaria [39]. *Millepora* spp. are members of the monogeneric family Milleporidae, the sub-order Capitata and the order Anthomedusae [40–42]. Milleporidae and Stylasteridae are the only two Hydrozoan families producing skeletons of calcium carbonate. The first scientific report of *Millepora* spp. was from Linnaeus in 1758 [43], with subsequent species reports and descriptions by several authors (e.g. [44, 45]), and the seminal work of Boschma [46, 47]. There has been a surge of interest in fire corals over the last two decades (**Figure 1**), and especially in the last 7 years. These recent studies focused on genetics and coincided with the development of new molecular markers [48–50]. While there is much known now about fire corals (reviewed in [51]), the literature is scattered, particularly that of biogeography and population genetic research, and needs to be summarized. In this chapter, we will document what is known about taxonomy, biogeography, ecology, symbiosis, morphology and reproductive biology of *Millepora* hydrocorals, using both published and unpublished information, and will highlight areas where knowledge is especially lacking.

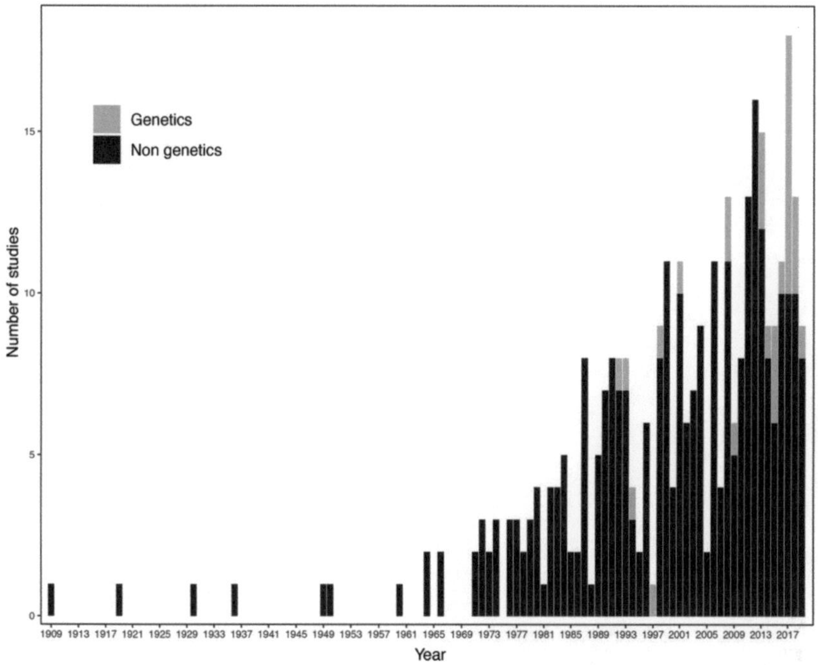

Figure 1.
The literature search in web of science identified 326 publications referring to Millepora *hydrocorals, wherein only 29 were using genetic approaches.*

2. Biodiversity and biogeography

2.1 Fossil records

Hydrocorals have a relatively long evolutionary history since many fossils from Tertiary deposits have been assigned to *Millepora*. However, Boschma [52] recognized only two species, *Millepora tornquisti* from Eocene rocks (56–33 mA) of Madagascar and *M. alcicornis* in Pleistocene deposits (2.58 mA–11,700 ya) from the Panama Canal zone [53]. Other branching milleporids were also reported from the Upper Cretaceous (100–66 mA) in northern Spain [54]. Recently, fossils of *M. alcicornis* have been recorded in deposits from the Early Miocene (∼23 mA) [55]. *M. exaesa* fossils were also recorded in more recent deposits in the Seychelles, dating from the last interglacial sea-level high-stand, ∼129,000–116,000 ya [56].

2.2 Species delimitation

As in many corals, the morphological species concept was traditionally applied to the species delimitation of *Millepora*, which is based on colony growth forms. *Millepora* species have a great diversity of growth forms and can be encrusting, branching, plate-like, massive or even columnar (**Figure 2**). Interestingly, the typical growth forms of *Millepora* species are broadly the same in the Red Sea and the Indo-Pacific. Arrigoni and colleagues [57] hypothesized a morphological convergence for these species. Similarly, in the Atlantic, there is also one plate-like and one branching species, as well as other massive/encrusting forms. As these growth forms do not form monophyletic groups on the phylogenetic reconstructions [57],

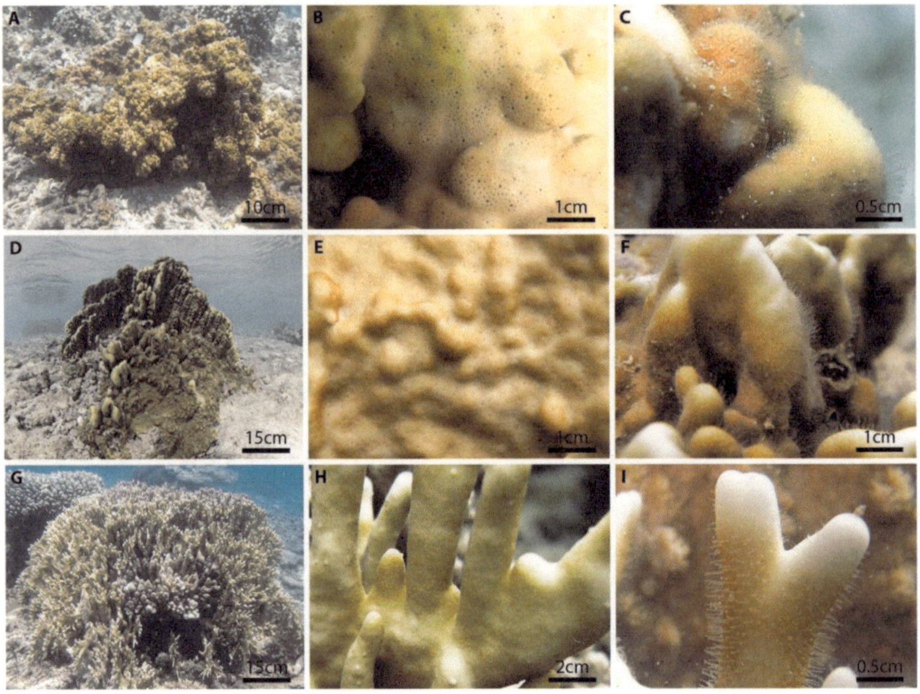

Figure 2.
Growth forms, pores and polyps of three Millepora *species. (A–C) M. cf. exaesa encrusting growth form, pores and polyps, respectively; (D–F) for the massive M. cf.* platyphylla *and (G–I) for the branching M. cf.* tenera. *D photograph is courtesy of Gilles Siu.*

they seem to have appeared independently and likely evolved in relation to the hydrodynamic conditions of their environment.

This group is also known for its great phenotypic plasticity [46], and environmental factors are known to greatly influence the morphology of *Millepora* colonies. Recently, Dubé and colleagues [58] demonstrated phenotypic plasticity among clonal colonies distributed in habitats with different hydrodynamic characteristics (see Section 4). To further complicate the matter, fire corals have been shown to overgrow stony corals, hydrocorals and gorgonians, which gives them additional peculiar growth forms (**Figure 3**) [59, 60]. Consequently, about 100 nominal species were described [61]. While Duchassaing and Michelotti [44, 62] identified 24 *Millepora* species based on trivial morphological differences, Hickson [45, 63] reckoned that there was only one *Millepora* species, *M. alcicornis*, and that all other morphological growth forms were only ecological variations. There is a true 'species boundary problem' within *Millepora* and it has been subject to much debate for over 150 years [45–47, 64, 65].

Apart from colony growth forms, pore traits are the most widely used characters in *Millepora* species delimitation. The pores in *Millepora* are like the corallites for the scleractinian corals, accommodating the polyps. There are two types of polyps in *Millepora* species: feeding polyps (gastrozoids) which are provided with a gastrovascular cavity opening by a mouth, and defensive polyps (dactylozoids) without a mouth. The gastropores and the dactylopores, from which the gastrozoids and the dactylozoids are able to extend outside to catch food, are organized in cyclosystems formed by a circle of dactylopores surrounding a single gastropore (**Figure 2**). While Boschma [46] concluded that the colony growth form was the most important character for the distinction of species, and that the other characters were not sufficient delimiting criteria, subsequent studies have used pore

Figure 3.
Millepora *hydrocorals overgrowing living reef corals at Europa Island (Indian Ocean), including massive Porites (A, C and D), Distichopora sp. (B) and Astrea sp. (D). M. cf.* platyphylla *can overgrow giant clam shells (C).*

characters. Using a more elaborate quantitative approach on pore characters, Moschenko [66] considered 11 traits (e.g. numbers and diameters of gastropores and dactylopores, distances between dactylopores and gastropores, number of dactylopores per gastropore) in one plate-like (*M. platyphylla* Hemprich and Ehrenberg 1834) and five branching *Millepora* species (*M. cruzi* Nemenzo 1975, *M. dichotoma* Forskal 1775, *M. intricata* Milne Edwards 1860, *M. murrayi* Quelch 1884 and *M. tenera* Boschma 1949). His results distinguished only *M. platyphylla*, while all branching species shared important overlap in trait values with gradual transition from one species to another [66]. However, *M. cruzi* and *M. murrayi* have been subsequently synonymized (with *M. tenera* and *M. intricata*, respectively) and this could explain some of the trait overlaps between species. More recently, Razak and Hoeksema [65], based on colony growth forms and pore characters, revised the Indonesian *Millepora* species and synonymized 6 of the 13 recognized Indo-Pacific species. In particular, the gastropore and dactylopore diameters were shown to be discriminant among many *Millepora* species [57, 67, 68]. Boissin and colleagues (submitted) analyzed 13 pore characters and could distinguish the three species present in Reunion Island. This latest study showed that gastropore and dactylopore numbers, as well as diameters, were informative and should be used as standard traits in future *Millepora* studies. This study also showed that polyp features were discriminant, such as the presence or absence of capitate tentacles or capitations, and the presence, absence or abundance of Symbiodiniaceae. Additional biological traits seem to be helpful to delineate milleporid species, such as reproductive periods, medusoid features and nematocyst morphology [36, 57, 69, 70].

The advent of DNA barcoding greatly helped delimiting species of many marine invertebrates [71–74]. Consequently, the more recent works on *Millepora* spp. used a combination of morpho and molecular characterization. Mitochondrial sequence data were successfully used to delineate milleporid species from the Caribbean, revealing two genetic entities: *M. squarrosa* Lamarck 1816 and a species complex composed of *M. alcicornis* Linnaeus 1758–*M. complanata* Lamarck 1816 [67]. Similarly, the four *Millepora* species from the Brazilian province were discriminated using the 16S mitochondrial gene coupled with morphological characters [68]. Recently, a study on milleporids from the Red Sea successfully distinguished three species *M. platyphylla*, *M. dichotoma* and *M. exaesa* Forskal 1775, using both morphological and molecular characterization [57]. Similarly, Boissin and colleagues (submitted) successfully used 16S sequences to delineate the three *Millepora* species from Reunion Island.

2.3 Biogeography

Fire corals are found in tropical/subtropical regions around the globe, nearly ubiquitous on reefs in the Atlantic, Indian and Pacific Oceans (**Figure 4**). Currently, 10 species are considered valid in the Indo-Pacific and 6 in the Atlantic Ocean [57, 61, 65, 68, 75]. The species status of two other Indo-Pacific species, *M. nodulosa* Nemenzo 1984 and *M. latifolia* Boschma 1948, are still unclear [65]. Several Indo-Pacific species show an extensive geographic distribution from west of the Indian Ocean to west (*M. dichotoma*, *M. tenera*), centre (*M. platyphylla*) or east of the Pacific Ocean (*M. exaesa, M. intricata*), while *M. foveolata* Crossland 1952 and *M. boschmai* de Weerdt and Glynn 1991 have restricted distributions (Philippines and Indonesia, respectively, **Figure 4**). In the Atlantic, two species are endemic to the Caribbean province (*M. complanata*, *M. squarrosa*) and three species are endemic to the Brazilian province (*M. braziliensis* Verrill 1868, *M. nitida* Verrill 1868, *M. laboreli* Amaral 2008), while *M. alcicornis* is present in both provinces as well as in the Canary Islands, Cape Verde and Ascension Island (**Figure 4**) [46, 76].

However, with recent morpho-molecular re-evaluations of species boundaries in this group, our understanding of the biogeographic patterns is still evolving. The recent highlight of cryptic species between the Red Sea and the rest of the Indo-Pacific provinces [57] pointed out that *M. platyphylla*, *M. dichotoma* and *M. exaesa* in the Indo-Pacific need taxonomic re-description. The number of Indo-Pacific species was thus raised from 7 to 10 in the last few months. This number is likely to grow in future years, as *M.* cf. *exaesa* for instance includes several lineages over its Indo-Pacific range and likely represents another case of species complex (Boissin et al., unpublished).

Additionally, the range of *M. platyphylla* (now *M.* cf. *platyphylla*) was recently extended back to the eastern Pacific [77] from where it was documented as extirpated decades ago [78]. In the Atlantic, *M. alcicornis* has recently established in the Canary Islands (Macaronesia), far north of its tropical distribution [79], possibly by means of drifting material from the Caribbean Sea or transportation through ballast waters of large vessels and fouling of hulls [79–81]. Long-distance dispersals in milleporids have also been demonstrated in the Pacific, with *Millepora* colonies recorded on drifting pumice [80]. This alternative mode of dispersal can explain such a wide geographic distribution for a species with a short pelagic stage (see Section 5.3). However, as noticed by Lewis [51], it is surely remarkable that a family of worldwide distribution, with a long geological history and apparent ecological success, is represented by less than 20 species.

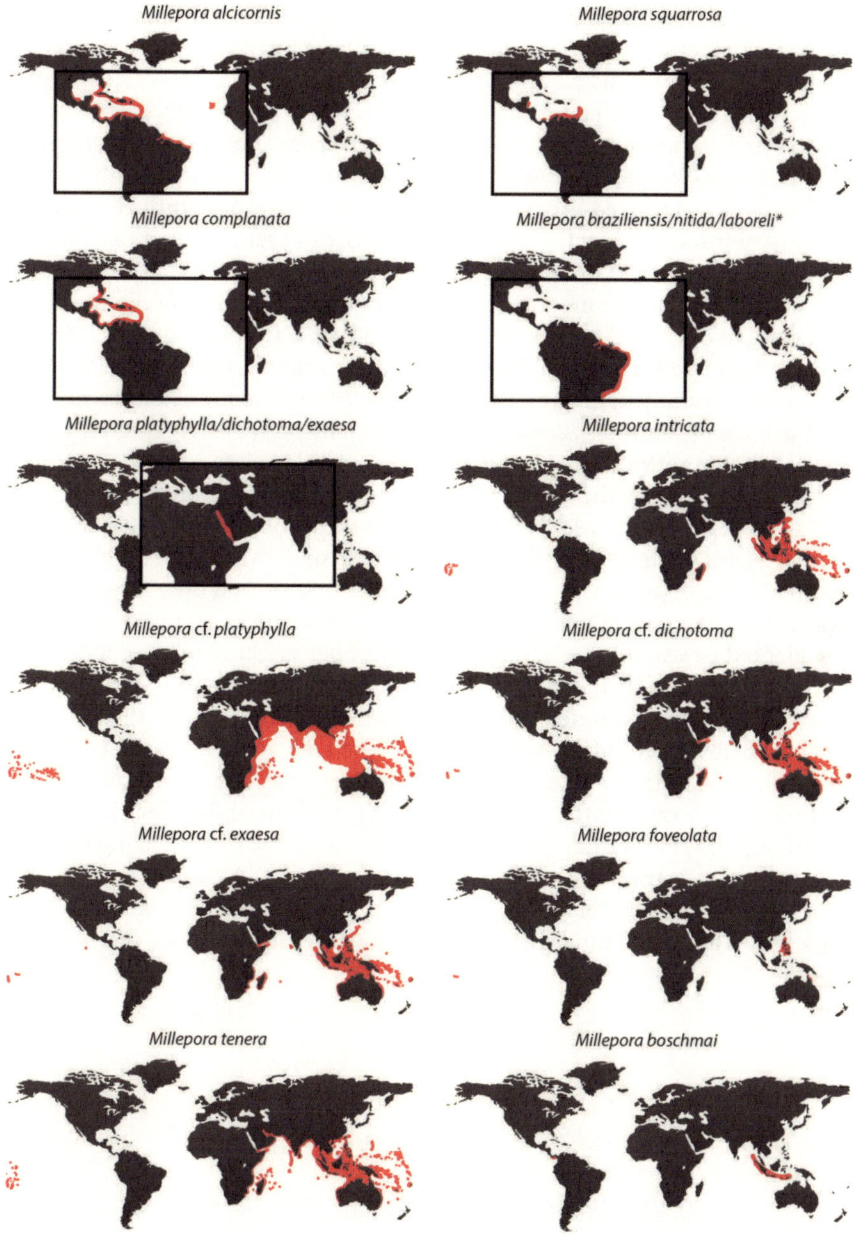

Figure 4.
Geographic distribution of the 16 recognized species of Millepora *in the Atlantic and Indo-Pacific Oceans.*

3. Ecology and symbiosis

3.1 Distribution, abundance and ecological roles

Fire corals occur worldwide in tropical seas and are limited in distribution from the intertidal zone to depths of approximately 50 m [51, 82, 83]. Although fire corals can be abundant locally [84–86] and dominate shallow water communities in some coral reef ecosystems [87–90], they usually cover less than 10% of the overall reef substratum [51, 91]. *Millepora* spp. are also found in many environments and waves, water movement, light intensity and habitat depth were identified as key factors influencing their distribution and growth forms [51, 82, 91–93]. On barrier reefs, the amount of wave energy is highest on the reef crest, where wave breaking first occurs and subsequently attenuates towards fore reef and lagoonal environments (**Figure 5**) [94, 95]. This gradient in wave energy, combined with *Millepora*'s sensitivity to wave-induced breakage, were showed to strongly influence colony and size distributions of *M.* cf. *platyphylla* at Moorea (French Polynesia), with highest densities recorded on the fore reef and larger colonies on nearshore reefs [91]. *M.* cf. *platyphylla* colonies occurred in a contagious pattern of distribution (i.e. colonies close to one another), as described for other Caribbean species [96], and colony breakage and subsequent fragment re-attachment were suggested as explanations for such colony aggregations [58]. Three *Millepora* species were also identified on the reefs of Reunion Island [97], where each species is distributed according to their proximity with the shore and reef crest, mostly related to the wave energy dispersal. *M.* cf. *exaes*a is the first species encountered close to shore on the shallow reef flat (2 m depth), replaced by *M. tenera* when approaching the reef crest, and *M.* cf. *platyphylla* colonies live from the crest to 35 m depth on the outer slope.

Millepores are important reef framework builders, second only after scleractinian hard corals [51, 82]. Their complex structure is a habitat for other species adapted to stinging cells, including scavenger crustaceans (e.g. crabs, shrimps and barnacles, [51, 98–100]), as well as fish [38, 101–103], serpulids [104, 105], spionid polychaetes [51] and scleractinian corals [106]. Interestingly, high fire coral cover on Caribbean reefs was associated with increased fish richness species [86]. Many studies have described hydrocorals as opportunistic species that show rapid growth rates with high fecundity [51] and the ability for clonal propagation through fragmentation [58]. Fire corals are capable of colonizing both natural and artificial substrates, including dead gorgonians, rocks and ships [107, 108], as well as living seagrass stems, hydrocorals, gorgonians, scleractinian corals and other reef invertebrates (e.g. giant clams) through pursuit, contact and overgrowth

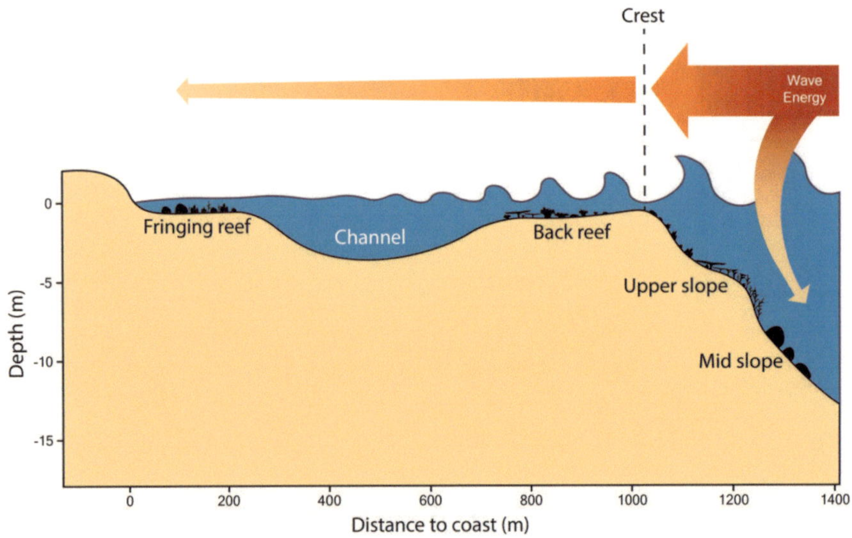

Figure 5.
Wave energy dispersal on a barrier reef (modified from [94, 95]). The fore reef experiences strong wave action from incoming waves that break on the reef crest, near the upper slope, with a significant decrease in swell exposure towards deeper waters. The reef crest dissipates ~70% of the incident swell wave energy with gradual wave attenuation from the back reef to nearshore fringing reefs.

(**Figure 3**) [59, 60]. This ability to inhabit different substrates and its rapid colonization rates [79] provide a competitive advantage for potential habitat expansions. Although fire corals compete with other corals, they also contribute to coral survival during *Acanthaster* outbreaks [106], highlighting their key ecological role in reef resilience. In fact, the corallivorous predator *Acanthaster planci* tends to avoid *Millepora* species [109], thus providing predator-free sanctuaries to nearby scleractinian corals.

3.2 Endosymbiosis with photosynthetic dinoflagellates (Symbiodiniaceae)

Many members of the phylum Cnidaria, including corals, octocorals, sea anemones and hydrocorals, host unicellular dinoflagellate endosymbionts (i.e. zooxanthellae) belonging to the family Symbiodiniaceae [110]. These associations are often obligatory and of fundamental importance to coral reef ecosystems as they enhance the growth of calcifying corals that form the reef. For instance, the zooxanthellae contribute to host nutrition (up to 95% of the energy requirements in scleractinian corals [111]) and skeletogenesis by providing photosynthetically fixed carbon, while the cnidarian host provides inorganic nutrients and refuge from herbivory to its symbionts [112–114]. Previous studies have demonstrated that the association of Cnidaria–Symbiodiniaceae is not stochastic, but mostly determined by host phylogeny and geography [115, 116]. Like scleractinian corals, hydrocorals feed heterotrophically on a variety of resources (mostly planktonic feeders [51, 117]) and rely on a mutualistic symbiosis with Symbiodiniaceae algae for autotrophic nutrition and calcification [118, 119]. While coral-Symbiodiniaceae associations have been extensively studied over the last decades (reviewed in [120]), only two studies have recently investigated hydrocoral-Symbiodiniaceae associations on Caribbean reefs [121, 122]. Rodriguez and colleagues [122] showed that Symbiodiniaceae species that associate with *M. alcicornis* vary as a function of its geography, with *Symbiodinium* sp. (formerly clade A) found in samples from Mexico and *Breviolum* sp. (formerly clade B) in the eastern Atlantic, with the exception of samples from the Canary Island and Cape Verde Islands that comprised *Cladocopium* sp. (formerly clade C). Unpublished data collected across *M.* cf. *platyphylla* Indo-Pacific range showed that this species can associate with the genera *Symbiodinium* (dominant symbiont), *Cladocopium* and more rarely with *Brevolium* in French Polynesia, Papua New Guinea and the south-western Indian Ocean (Dubé et al. in prep.; Boissin et al. in prep.). The other Indo-Pacific species (*M.* cf. *dichotoma* and *M.* cf. *exaesa*) investigated so far show the same Symbiodiniaceae associations (Boissin et al. in prep.).

3.3 Bleaching susceptibility

One of the most devastating consequences of global warming is coral bleaching. Bleaching occurs when scleractinian corals, hydrocorals and octocorals lose their photosynthetic symbiotic algae or pigments [21, 111, 123–125], which leads to the white calcium carbonate skeleton being visible through the transparent host tissue. The frequency and severity with which coral bleaching occurs have increased in recent years [126]. Numerous investigations have demonstrated that coral bleaching events are a serious threat to coral reefs worldwide, where they have caused a severe deterioration in reef health (e.g. increase in coral disease, decrease in reef calcification and loss of habitat for related reef organisms [25, 123, 127–129]. The severity of coral bleaching depends on several factors, including specific coral species impacted [130], symbiotic algae assemblages [131] and thermal history [132].

Zooxanthellate hydrocorals are thought to be extremely sensitive to bleaching [130, 133] and can be threatened by future climate change. *Millepora* spp. have been reported to be among the first cnidarians to lose their zooxanthellae symbionts during widespread bleaching events [134] and they have suffered local or regional extinctions from bleaching in the Pacific [78, 85, 135]. Numerous investigations of bleaching events on Caribbean and Florida Keys reefs have reported bleaching of *Millepora* colonies [133, 136–139], with *M. alcicornis*, a finely branched species, being the most severely affected reef corals. Such coral morphology has been described to be more susceptible to bleaching than encrusting and massive species [140]. Yet, bleached colonies of *M. alcicornis* remained alive during a bleaching event affecting a north-eastern Brazilian reef [133], which is in accordance with previous reports that *Millepora* species are also the first to recover from short-term bleaching [136, 137]. In the Maldives Archipelago (Indian Ocean), *Millepora* was reported to be the major reef-building coral in shallow reefs (7 m depth), producing some '*Millepora* zones' [141]. Three species were well documented, the massive species *M.* cf. *platyphylla* [46, 142] and two branching ones, *M. tenera* [51, 143, 144] and *M. latifolia* [143]. However, many recent surveys of the Maldivian reefs have identified another pattern of distribution, where none to low abundances of *Millepora* species were recorded (1–2 depending on the species) [145–148]. Gravier-Bonnet and Bourmaud [148] suggested that milleporids were extirpated from several Maldives atolls, following the 1997–1998 El-Nino Southern Oscillation event (ENSO). ENSO has induced a strong bleaching and massive coral mortality (of up to 90%) in the tropical Indian Ocean, including the Maldives [145, 149]. On the Great Barrier Reef, *Millepora* spp. were also the most susceptible taxa to the mass bleaching event of 1998, with 85% of mortality [130], while they showed no evidence of bleaching at Moorea, although scleractinian corals were severely bleached at this location [150]. During 2014–2017, the worst documented bleaching event observed [26, 27], *M.* cf. *platyphylla* showed no sign of bleaching at Moorea, although about 60% of scleractinian corals were bleached on the fore reefs (**Figure 6A**). Since February 2019, Moorea's reefs are suffering from another mass bleaching event, with colonies of *M.* cf. *platyphylla* showing sign of bleaching and mortality (**Figure 6B**). Differential susceptibilities to this bleaching event were also observed between *M.* cf. *platyphylla* colonies (**Figure 6C**). Ongoing surveys will help quantifying bleaching susceptibility and mortality among coral taxa and locations, as well as between fire coral growth forms and genotypes (Dubé et al. in prep). Nevertheless, a previous study has shown that temperature is the primary factor related to bleaching in *M. alcicornis*, but that synergism with exposure to solar radiation may play a key role in hydrocoral bleaching [151]. Also, multifocal bleaching in hydrocorals, consisting of numerous scattered bleached spots, has been first described as a syndrome caused by an infectious disease affecting several colonies of *M. dichotoma* in the Red Sea [152]. 16S rRNA gene sequencing showed that affected tissues match sequences of bacteria belonging to *Alphaproteobacteria* and *Bacteroidetes* members previously associated with various diseases in scleractinian corals [153]. Yet the mechanisms of multifocal bleaching, its aetiology and mode of transmission remain unknown. Nevertheless, many studies have addressed the aetiology and effects of bleaching in Anthozoan species, wherein changes in the expression of genes and proteins were observed, and particularly heat shock proteins and transcription factors [154–159]. A recent study demonstrated that bleached specimens of *M. alcicornis* in Mexican Caribbean undergo a moderate decrease in symbiont's density and photosynthetic pigments, in addition to differential expression of 17 key proteins, such as calmodulin, actin and collagen

Figure 6.
Bleaching susceptibility of M. cf. platyphylla *during massive bleaching events occurring on the fore reefs at Moorea Island (French Polynesia). (A) View of the fore reef at Moorea during the bleaching event of 2016, showing healthy colonies of M. cf.* platyphylla *and bleached colonies of scleractinian corals, mostly of the* Pocillopora *genus. M. cf.* platyphylla *was sensitive to the recent bleaching event of 2019 at Moorea, where colonies bleached and died (B) from the rise in temperature, while other colonies showed sign of resistance to bleaching on the same reef (C). Photographs are courtesy of Yannick Chancerelle (A) and Yann Lacube (B and C).*

often coupled with calcium homeostasis, exocytosis and cytoskeleton organization in Anthozoan species [139].

Coral reefs are also threatened by ocean acidification associated with the increasing CO_2 partial pressure, which depresses net calcification of corals and hydrocorals [160, 161]. Physiological responses of reef organisms to ocean acidification are relatively well known [162, 163]. Examples include changes in gene expression consistent with metabolic suppression, increased oxidative stress, antioxidant system, apoptosis and symbiont loss [164, 165]. Yet little information on the effects of ocean acidification on the physiology of fire corals is available in the current literature. Luz and colleagues [166] demonstrated that the antioxidant defense system of *M. alcicornis* is capable of coping with acidic conditions for a short period of time, while long-term exposure induces oxidative stress with consequent oxidative damage to lipids and proteins, which could compromise hydrocoral health and influence negatively the zooxanthellae-coral symbiosis and ultimately lead to bleaching [167].

4. Morphology and phenotypic plasticity

In coral reefs, some calcifying species, such as corals and hydrocorals, are known to have a high degree of morphological plasticity in response to hydrodynamic changes and light availability, which strongly influences their performance, including resource acquisition and light capture, thereby benefiting colony growth, reproduction and survival [168]. Branching and plating growth forms grow quickly into large arborescent colonies in shallow reef environments, where irradiance is high and water flow is low, which makes them effective competitors for space [169, 170], light and food [171]. However, this growth strategy renders them extremely vulnerable to breakage when large waves and storm events occur, often resulting in fragmentation or coral mortality [172, 173]. Intraspecific morphological variation has been reported in many colonial reef organisms in response to environmental gradients, which ultimately affect their survival and growth [174–177]. Such plastic developmental responses are often induced during ontogeny of modular organisms with persistent effect on adult phenotypes [178]. These phenotypic responses can also change independently from the genetic background of reef corals (acclimatization), but they often rely on a genetic basis (adaptation) [179, 180].

Fire coral species are also known for their extensive morphological variability and vulnerability to fragmentation varies greatly among their morphologies [51, 58, 91, 181]. Examples include variations in growth forms of *M.* cf. *platyphylla* colonies that were found in distinct reef environments at Moorea; the fore reef at 15 and 6 m depth (mid and upper slope, respectively), the back and fringing reefs [58, 91]. Colonies on the mid slope and back reef were mostly encrusting, while the massive morphology was dominant in the fringing and patch reefs (**Figure 7A, B**). The sheet tree morphology of *M.* cf. *platyphylla* [182], the most vulnerable to wave-induced breakage, was nearly exclusive to colonies encountered in the upper slope (**Figure 7C**), where waves can break the blades, while the encrusting bases remain intact [181].

To date, the flexibility for a single genotype to produce a range of phenotypic responses to distinct environmental conditions (i.e. phenotypic plasticity) has rarely been documented in natural marine populations, mostly because of the difficulty in identifying naturally occurring clonal genotypes across variable environments. Dubé and colleagues [58] have described the first example of phenotypic plasticity among fire coral clones, where clones of the same genotype display different morphologies across distinct reef habitats (**Figure 8**). The fire coral *M.* cf. *platyphylla* seems to invest in a vulnerable morphology that increases the contribution of asexual reproduction through fragmentation in high-energy reef habitats. This is a unique example of phenotypic plasticity as corals typically have wave-tolerant growth forms in such dynamic reefs. Such phenotypic responses suggest

Figure 7.
Morphologies of M. cf. platyphylla *colonies in habitats experiencing contrasting hydrodynamic regimes. (A) Massive wave-tolerant morphology in the patch reef, a lagoonal habitat (photograph is courtesy of Gilles Siu); (B) encrusting wave-tolerant morphology in the back reef, a lagoonal habitat at <1 m depth and (C) sheet tree morphology vulnerable to wave-induced breakage in the upper slope, a fore reef habitat at 6 m.*

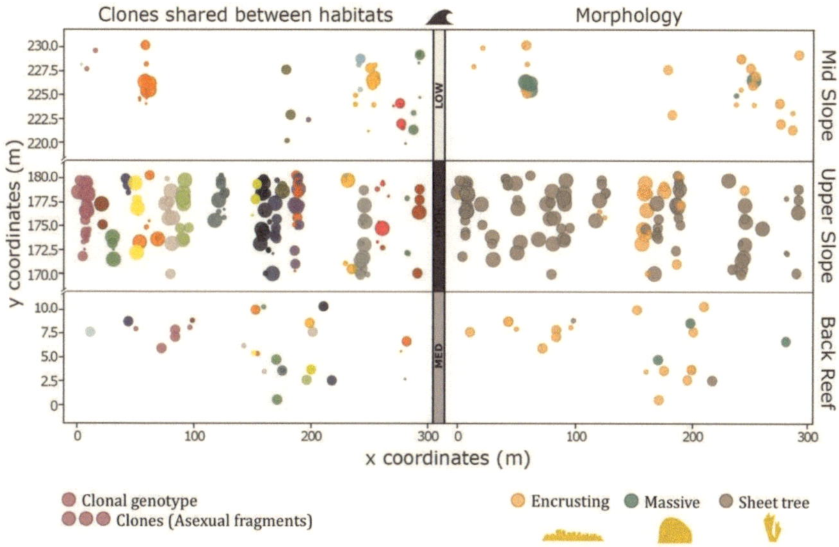

Figure 8.
Graphical abstract showing the occurrence of phenotypic plasticity among fire coral clones, where clones of the same genotype display different morphologies across distinct reef habitats [58]. Geographic coordinates of each georeferenced colony collected in the three reef habitats are shown in meters on the x and y axes. On the left side: each genotype is represented by a unique color; on the right side: colonies with encrusting morphology are shown in orange, massive in green and the vulnerable sheet tree morphology in grey.

that fire corals being susceptible to wave-induced breakage have benefits in terms of reproduction outweighing the costs of getting injured.

5. Reproduction

5.1 Reproductive strategies

Although only a few species are exclusively reproducing asexually, clonality has evolved repeatedly in many reef organisms (e.g. [183–186]). In coral reef ecosystems, there are many organisms that can reproduce through both sexual and asexual reproduction, including scleractinian corals [187], hydrocorals [58], hydroids [188], coralline algae [189], sea anemones [190], sea cucumbers [191], gorgonians [192] and sponges [193]. Asexual reproduction produces genetically identical offspring, often leading in local populations dominated by few adapted clones [194–196]. In the contrary, sexual reproduction enables genetic recombination and production of genetically diverse propagules, thus generating the genotypic variation required for adaptation [197] and colonization of new habitats [198]. In many colonial reef organisms, asexual reproduction can occur through fragmentation, fission, budding, polyp expulsion or polyp bail-out [187, 199–201], while sexual reproduction often involves a wide range of reproductive strategies, i.e. gonochorism, hermaphroditism, internal (brooders) and external (spawners) fertilization [187, 202].

Despite their ecological importance to the ecosystem functioning of coral reefs, *Millepora* hydrocorals have been relatively understudied and information regarding their reproduction and dispersal patterns remain scarce. Fire corals are gonochoric broadcast spawners that reproduce sexually by producing medusoids and planula larvae (**Figure 9**). They also rely on asexual reproduction through fragmentation [58, 181], but the production of asexual larvae has never been documented within this genus though described for some *Pocillopora* species [203, 204].

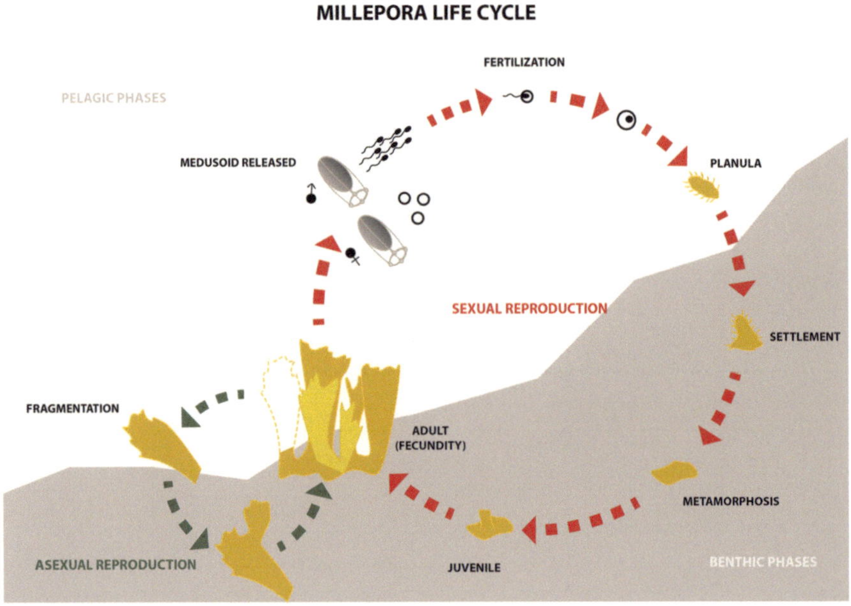

Figure 9.
Millepora *life cycle.* Millepora *hydrocorals are gonochoric broadcast spawners that reproduce sexually by producing medusoids and planula larvae. The medusoids release the gametes in the water column for external fertilization. The ciliate larvae sink and crawl on the reef substratum and metamorphose in a new calcifying polyp, founder of a new colony.* Millepora *also relies on clonal propagation through fragmentation and grow via asexual budding.*

5.2 Spawning, medusoids and larval development

Milleporid sexual reproduction is seasonal [69]. *Millepora* colonies become mature during the spring or summer (or austral summer for the southern hemisphere). Sexual reproduction period is usually correlated with the increase of the sea water temperature [69, 70, 205, 206], but some studies based on ampullae observations suggest a reproduction throughout the year [207–209]. Spawning occurs at different dates according to species, preventing hybridization [69, 70, 206]. The empty ampullae are visible during 1–2 months on the colonies (**Figure 10D**) before the skeleton reconstruction.

The sexual reproduction process begins with the growing of special cavities, called ampullae, developed in tissues and designing densely packed white dots on the coenosteum of the gonochoric colonies. These ampullae were first described by Quelch [210, 211] and further studied by Boschma [46, 207, 208, 212] and Moschencko [66]. Each ampulla contains one developing medusoid, i.e. a 'regressed' short-lived medusa, shed with mature gametes. Male and female medusoids are liberated after the disintegration of the dense network of the trabeculae covering the ampullae (**Figure 10A–C**). They have marginal bulbs but no tentacle, no circular or radial canal, no manubrium, no statocyst or any sense organ (**Figures 10C** and **11A**), and they are not able to feed on zooplankton. On the contrary, as true medusae, they are able to actively swim with their muscle fibres distributed in the bell and display a velum. Gonads are attached to the short spadix and fill entirely the subumbrellar cavity. Female medusoids contain 2–5 zooxanthellate oocytes (**Figure 11A**) and male medusoids contain a spermatic mass (**Figure 10C**). The medusoids detach themselves from the fertile colonies by active bell pulsations in a few minutes and their swimming activity leads to the release of

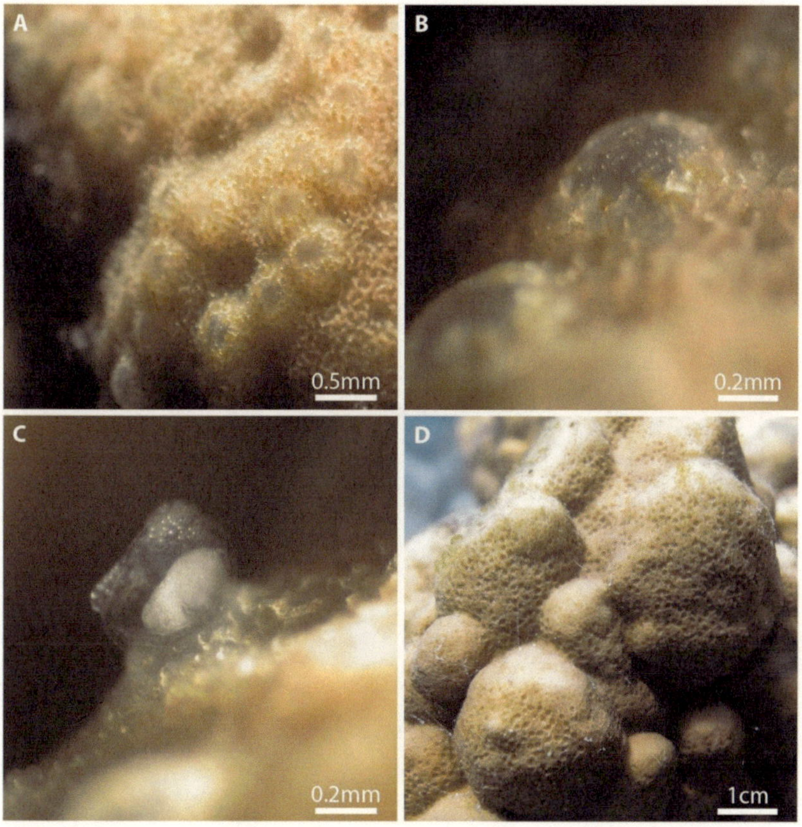

Figure 10.
Before, during and after medusoid release in M. cf. exaesa *in Reunion Island (modified from [69]). (A) Ampullae showing a small opening resulting in skeleton dissolution few days before the medusoid release. Notice that the cyclosystems have disappeared because of the high ampulla density; (B) medusoids protruding through the open ampullae and (C) male medusoid release with the umbrella opening towards the surface. Notice the big tentacular bulbs with refringent nematocyst and the sperm sac filling the subumbrellar cavity; (D) empty ampullae visible during 1–2 months after the massive medusoid release event. A, B and C photographs were taken using a stereomicroscope; photograph D was taken underwater.*

the ripe gametes in the water column. The spawning of gametes is therefore almost synchronous with the release of medusoids. Spawning always begins before dark, but is not correlated with the lunar or tidal cycles [69, 70, 206]. In shallow water of Reunion Island, Indian Ocean (reef flat), a unique massive spawning event was observed *in situ* for *M.* cf. *exaesa* and *M.* cf. *platyphylla* during the reproductive period, in December for the former species and in January for the later one [69]. Conversely, *M. tenera* seems to spawn regularly but not massively during 2 months of the austral summer, resulting in the observation of both closed and open ampullae on fertile colonies during the reproduction season. Likewise, Nomura [205] and Soong and Cho [206] described several medusoid batches in different *Millepora* species in controlled conditions during the reproductive season in Japan and Taiwan, respectively. Recently, Shlesinger and Loya [70] described massive spawning events in the Red Sea (Gulf of Eilat/Aqaba) for three species, *M. dichotoma*, *M. exaesa* and *M. platyphylla*. Their field observations during the reproductive period (from June to September 2016–2018) also showed one or two spawning events per year for *M. exaesa* and *M. platyphylla*, while *M. dichotoma* colonies released their medusae massively, 4–6 times during the reproductive season. The higher reproductive output of *M. dichotoma* might be in relation with its

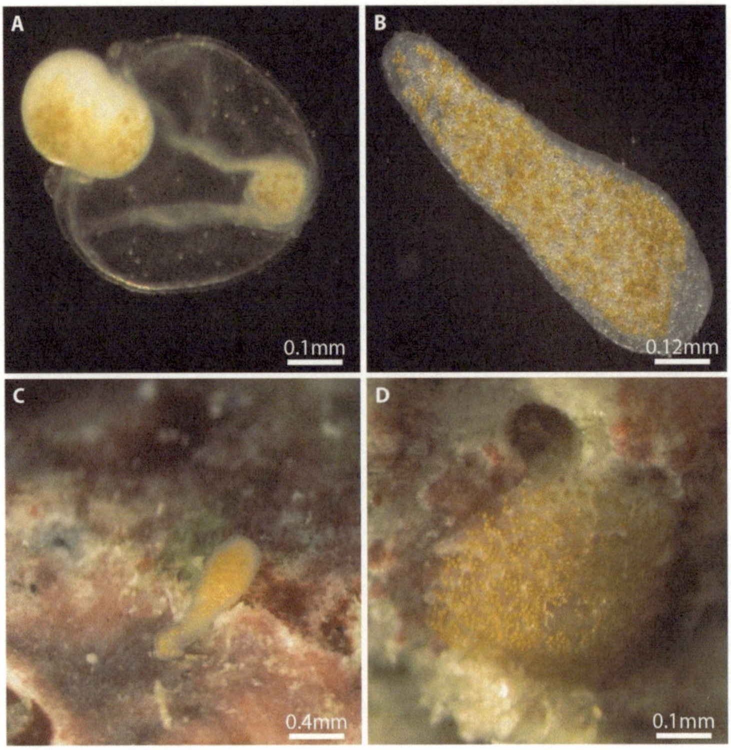

Figure 11.
Gamete spawning, planula larva formation and settlement in Millepora *spp. in Reunion Island (modified from [69]). (A) M. cf.* platyphylla *female medusoid releasing an oocyte through the velum while swimming. Notice the numerous zooxanthellae in the oocyte and spadix tissues; (B) M. cf.* exaesa *zooxanthellate (orange dots in endoderm) planula larva; (C) M. cf.* exaesa *larva finding a sustainable substrate to fix by the tapered pole before metamorphosis; (D) M. cf.* exaesa *recruit with the first pore. All photographs were taken using a stereomicroscope.*

higher abundance in the Gulf of Eilat/Aqaba (i.e. *M. dichotoma* is the most abundant milleporid in the Gulf [70]).

The empty medusoids continue to swim for 1–3 h and die quickly while sinking and shrinking. Male and female medusoids are released synchronously (for a giving species), the spawning of the oocytes and spermatozoids is also simultaneous, and fertilization occurs rapidly. Embryogenesis and planula larvae formation occur in less than 12 h in aquarium [69]. Because of the presence of algal symbionts in oocytes, the planula larvae are zooxanthellate and have the potential to live for several weeks before settlement (more than 1 month in controlled conditions for *M.* cf. *exaesa* from Reunion Island). This feature is certainly a character to keep in mind to explain the large distribution of *Millepora* species in all oceans. *M.* cf. *exaesa* planula has been described as a bipolar ciliated larva with a wide anterior and tapered posterior, without a mouth and gastrovascular cavity (**Figure 11B**) [69]. The larva endoderm is full of lipid droplets and zooxanthellae. The larva sinks and crawls until it finds a sustainable substrate to fix and metamorphose (**Figure 11C**). This process leads to the formation of a calcareous structure surrounding the primary polyp, founder of a new colony by asexual budding (**Figure 11D**).

The reproductive output (ampulla density) is variable according to species and within species. Amaral and colleagues [75] found an average of 10 ampullae/cm^2 for *Millepora* species occurring on Brazilian reefs, while the highest density was observed by Soong and Cho [206] in Taiwan with 84–120 ampullae/cm^2.

In Hydrozoans, the reproductive output can vary between and within species, and can often depend on the colony size and environmental conditions [213, 214]. In Reunion Island, the ampulla density of *M.* cf. *exaesa* is positively correlated with the size of colonies, indicating that the reproductive output varies with the colony size. Global change also seems to influence the reproductive output of milleporids as the rate of fertile colonies have decreased considerably in the last 10 years at two contrasting reef sites in Reunion Island (Bourmaud et al. in prep).

5.3 Dispersal and recruitment

For most colonial reef species whose adults are sessile, their early life history includes a pelagic stage. These propagules represent the first step for successful recruitment and have profound implications for population dynamics and renewal, which ultimately affect their evolutionary history [215, 216]. Dispersal in colonial organisms is mostly mediated by the release of gametes and/or larvae during sexual reproduction events, together with the continuous supply in asexual propagules. In many reef species, the extent of dispersal is largely governed by the reproductive biology and early life history ecology. Molecular studies and oceanographic models have uncovered a wide range of dispersal patterns (i.e. dispersal kernels) in coral reefs, from populations primarily sustained by self-recruitment due to limited dispersal potential or retention, to ecologically significant gene flow and connectivity among adult populations [217]. In corals for instance, brooded larvae settle and metamorphose rapidly after being released, which is most likely to enhance local dispersal patterns, while broadcast larvae require a planktonic development phase and settle further away from the parental source [187]. On the other hand, clonal propagation can allow populations to expand locally under unfavorable conditions. Such conditions include fragmented [218], marginal [196] and highly disturbed habitats [186], where clonal reproduction reinforce local adaptation processes and population genetic heterogeneity due to restricted dispersal potential of asexual offspring [58, 219, 220].

Although local demography and self-recruitment have been shown to have major consequences on the genetic diversity and adaptive ability of reef organisms, empirical data of dispersal patterns in reef-building species remain scarce. Dubé and colleagues [221] documented the first genetic estimates of local dispersal and self-recruitment in a marine broadcasting species, the hydrocoral *M.* cf. *platyphylla*. They performed a parentage analysis that revealed a significant contribution from self-recruitment in addition to limited dispersal of sexual propagules on Moorea's reefs. Sexual propagules often settled at less than 10 m from their parents and dispersal events decreased with increasing geographic distances. Sibship analysis showed that full siblings recruit together on the reef, resulting in sibling aggregations. Such limited dispersal abilities in fire corals can be related to their early life history traits. Dispersion during the medusoid stage may not be as effective due to the short pre-competency period time of the hydromedusae in the water column [51, 181]. Other means of dispersal can occur through the propagation of asexual offspring, e.g. fragments that have broken and re-attached to the reef framework. Asexual reproduction through fragmentation in branching hydrocoral can be substantial during disturbances [51, 181] and may therefore contribute to dispersal. However, clonal fragments of the plate-like *M.* cf. *platyphylla* were found to be dispersed close to one another on a barrier reef (mean = 18 m), with clone distribution being perfectly aligned with wave energy dispersal [58]. The maximal distance between fragments of the same genotype in this plate-like species at Moorea Island was about 450 m.

6. Modularity and growth

Modularity is a well-established life history strategy among colonial reef invertebrates, i.e. corals, gorgonians, sea anemones, hydroids, hydrocorals, bryozoans and sponges [222]. Modular organisms grow in size via the repeated, vegetative formation of genetically identical modules, referred to as asexual budding, whereby all modules are derived from the same initial zygote to form a colony [223, 224]. Colony size often correlates with many fitness advantages in response to both physical and biological stressors. For instance, larger colonies can survive better towards predation [225] and competition [226], and their fecundity is often increased due to the large number of polyps that contributes to sexual reproduction [227]. Modules usually remain physiologically interconnected, but may also separate from the colony through fission or fragmentation and persist as discrete units [228], thereafter reducing colony size. There are only few reports of growth rates in *Millepora* species [79, 92, 229–232] that are within the range reported in Acroporidae corals from western Atlantic region [233].

Some marine modular organisms, e.g. corals and ascidians, can also grow larger and quicker via the fusion of distinct colonies [178], which results in genetically heterogeneous colony, also referred to chimera. In addition to chimerism, somatic mutations may arise within a colony, which also results in intracolonial genotypic variability. Both chimerism (fusion) and mosaicism (somatic mutation) were identified in fire corals [234, 235]. At Moorea, for instance, fusion between siblings is likely to occur as fire corals have limited dispersal abilities and are often aggregated due to the co-settlement of their larvae [221]. Puill-Stephan and colleagues [236] demonstrated that high levels of relatedness between juvenile corals correlated with late maturation of allorecognition. The fusion of siblings could thus be related to a low conspecific acceptance threshold and/or a delay in allorecognition maturation for *Millepora* hydrocorals, as described in some hermatypic corals [237, 238]. Considering the common occurrence of somatic mutations in fire coral species, modularity might be a promising strategy to increase genotypic variability in populations that are predominantly sustained through asexual reproduction [235].

7. Population genetics: a case study of *Millepora cf. platyphylla* at Moorea, French Polynesia

Recent genetic studies have uncovered that geographically isolated populations, such as those of Moorea, appear to be more dependent on self-recruitment for local replenishment and sustainability [239, 240], highlighting the importance of studying local patterns of life history traits in keystone species. Moorea is a high volcanic island surrounded by a barrier reef with extensive fringing reefs and lagoon systems [241]. Lagoons and deep interrupted channels separate the fore reefs from the island, and the lagoon is connected to the oceanic waters via deep passes through the barrier reef. Furthermore, coral reefs surrounding Moorea Island have undergone a massive decline in coral cover from a recent outbreak of *Acanthaster planci* and cyclone *Oli* [242, 243], which provides a unique perspective from which to comprehend how fire corals can survive and recover from such disturbances.

By gathering genotypic and phenotypic data, Dubé and colleagues [58, 221, 235] were able to produce a complete picture of ecological and evolutionary strategies involved in the population persistence of *Millepora* hydrocorals. On Moorea's reefs, *M.* cf. *platyphylla* displays a wide range of strategies that ensure its survival by maximizing the acquisition of local resources. Self-recruitment and mosaicism successfully established diverse genotypes within *M.* cf. *platyphylla* population, while colony fragmentation contributed effectively to population growth (**Figure 12**), where a high number of clonal genotypes have the potential for phenotypic plasticity in response to environmental changes. Genetic data indicated that fragmentation is the dominant reproductive process generating the high abundance of fire corals at Moorea (80% of colonies were clones). Even small recruits were having multilocus genotypes identical to adults and were often positioned below the reef substratum, i.e. frequently on branches of dead coral colonies or side of crevices. These observations suggest that the successful recruitment of clones may be the result of clonal reproduction processes other than fragmentation, such as asexual planula larvae, because asexual fragments are less likely to re-attach on such inclined substrate. The release of ameiotic planula larvae was reported in a number of coral species [187], where larval behaviour allows the settlement of a new individual characterized by its mother genotype (clone mates). However, such clonal reproductive strategy has never been described for the *Millepora* genus, and requires further investigations. In Moorea, fire corals are sustained by a moderate degree of self-recruitment [221] suggesting that despite low gene flow, genetically diverse and locally adapted recruits can successfully establish high local population abundance via their subsequent growth, survival and fragmentation (as described in [244]). However, such populations are predicted to be vulnerable to severe disturbances owing to their isolation from potential source reefs and are often associated with increased extinction risks [245, 246]. A high potential for gene flow and connectivity has been revealed among islands of the Society Archipelago in French Polynesia for some scleractinian species (i.e. Moorea, Raiatea, Taha'a and Tahiti) [218]. Preliminary results from samples of *M.* cf. *platyphylla* collected in several islands from French Polynesia revealed significant genetic differentiation among archipelagos (Marquesas, Austral, Gambier, Society and Tuamotu, Boissin et al. unpublished), highlighting the importance of self-recruitment processes in population sustainability.

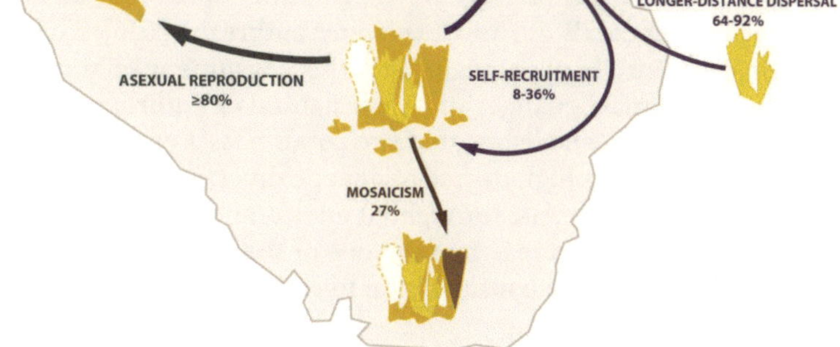

Figure 12.
Summary of life history strategies in M. cf. platyphylla *at Moorea, French Polynesia. M. cf.* platyphylla *heavily relies on asexual reproduction through fragmentation for local replenishment (80% of the colonies are clones), allowing population growth and the persistence of a genotype over time. M. cf.* platyphylla *population is sustained via a significant contribution from self-recruitment (8–36% of juveniles are self-recruits). Mosaicism and chimerism also contribute in creating novel genotypic diversity at the population and individual levels.*

Overall, the evaluation of the life history of *M.* cf. *platyphylla* suggests a competitive strategy, based on few locally produced sexual recruits and their ability of reaching large sizes (fusion [235] and stolonal spreading [59]), which allows them to pre-empt space on coral reefs, but also brought evidence of high susceptibility to fragmentation. This life strategy is well suited for population persistence in the absence of sexual recruitment, but can be risky in unstable environments [247]. Yet *M.* cf. *platyphylla* populations in Moorea have withstood severe disturbances, e.g. *Acanthaster* outbreaks, cyclones and mass bleaching events. Their recovery is foremost sustained by the rapid growth of remnant colonies, mostly those encrusting, and the subsequent local recruitment via both sexual and asexual reproduction. There is evidence that under pressure from environmental changes fire corals might be among the reef coral 'winners', joining some scleractinian species that have already been described as such [32, 85, 140]. Yet more information on how they respond to bleaching events is needed, as *Millepora* species have been reported to be highly vulnerable to thermal stress in other reefs [4, 130, 133]. Nevertheless, the life history of *M.* cf. *platyphylla* is most likely contributing to its colonization success in various reef environments in French Polynesia. Although *M.* cf. *platyphylla* is the only fire coral species reported in this geographic region [50], this species is also characterized by one of the widest ranges of distribution in the entire Indo-Pacific region within the *Millepora* genus [248], but similar to the branching species *M. intricata*. Evaluating the life history of other *Millepora* species with different growth forms will enable to determine whether these strategies are unique to *M.* cf. *platyphylla* or spread within the *Millepora* genus.

8. Conclusions

In recent decades, declines in scleractinian coral cover have challenged their role as key ecosystem engineers of coral reefs [25–27, 249–251]. Assuming rising sea temperatures and increased ocean acidification, climate change can interfere with a range of key processes in the life history of reef corals, including growth, calcification, development, reproduction and behavior [162, 252]. Despite the acclimatization and genetic adaptation of reef corals [2], such persistent physical and chemical conditions can lead to shifts in reef community composition. This phenomenon has already been reported in many reefs, where alternative organisms are dominating reef assemblages (reviewed in [253]). Only few studies have considered hydrocorals in ecological monitoring of coral reefs [130, 254, 255]. For instance, *M.* cf. *platyphylla* can dominate some reefs in the Indo-Pacific region [89] and also contribute to the survival of corals during *Acanthaster* outbreaks [106]. Therefore, it is crucial to gain insights into how populations of this keystone species can adapt and survive in the face of climate change, and other natural or anthropogenic disturbances. In this chapter, we established that fire corals possess a great variety of life history strategies that favor a high degree of genetic diversity and plasticity enabling these organisms to persist throughout environmental variations. Consequently, these *Millepora* species may become one of the major components in some modern reefs and requires more consideration in ecological monitoring.

Acknowledgements

Much of the information in this review has come from expeditions and field work we have made in the Indian and Pacific Oceans, courtesy of grants from the LabEx 'CORAIL', Regional Council of Reunion Island, ANR French funds, French

Biodiversity Federation (FRB), French Agency for Biodiversity (AFB), 'Observatoire des Sciences de l'Univers' from Reunion Island (OSU-R) and the Ministry of French overseas territories. We are very grateful to N Gravier-Bonnet who firstly initiated milleporids studies in the south-west Indian Ocean. Further, several collaborations lead us to undergo marine biodiversity surveys in the Indo-Pacific region. N Gravier-Bonnet also provided valuable help with data gathering and species descriptions. We would like to thank numerous colleagues that allow us to participate to biodiversity or multidisciplinary research programs (S Andréfouët, L Bigot, P Chabanet, JL Join) and many students for contributing to the studies. CED also wishes to thank M Ziegler, B Hume and CR Voolstra for helping in the identification of Symbiodiniaceae species associated with *M.* cf. *platyphylla* at Moorea and Reunion Islands. CED and EB were financially supported by a FRQNT PhD scholarship and Marie-Curie Postdoctoral fellowship (MC-CIG-618480), respectively, during expeditions, field work and data analysis.

Conflict of interest

The authors declare no conflict of interest.

Author details

Caroline E. Dubé[1,2]*, Chloé A.F. Bourmaud[2,3], Alexandre Mercière[1,2], Serge Planes[1,2] and Emilie Boissin[1,2]

1 PSL Research University, EPHE-UPVD-CNRS, USR 3278 CRIOBE, Université de Perpignan, Perpignan Cedex, France

2 Laboratoire d'Excellence "CORAIL", Moorea, French Polynesia

3 UMR ENTROPIE 9220, Université de La Réunion, Saint-Denis Cedex, La Réunion, France

*Address all correspondence to: caroline.dube.qc@gmail.com

References

[1] Veron JEN. Corals in Space and Time: The Biogeography and Evolution of the Scleractinia. Sydney: UNSW Press; 1995

[2] Hoegh-Guldberg O, Mumby PJ, Hooten AJ, Steneck RS, Greenfield P, Gomez E, et al. Coral reefs under rapid climate change and ocean acidification. Science. 2007;**318**(5857):1737-1742

[3] Srividhya S, Chellaram C. Role of marine life in nanomedicine. Indian Journal Innovations and Developments. 2012;**1**(S8):31-33

[4] Moritz C, Vii J, Lee Long W, Tamelander J, Thomassin A, Planes S. Status and Tends of Coral Reefs of the Pacific. France: Global Coral Reef Monitoring Network; 2018. 218 p

[5] Cesar H, Burke L, Pet-Soede L. The Economics of Worldwide Coral Reef Degradation. Cesar Environmental Economics Consulting: Arnhem; 2003

[6] Copper P. Ancient reef ecosystem expansion and collapse. Coral Reefs. 1994;**13**:3-11

[7] Spalding MD, Grenfell AM. New estimates of global and regional coral reef areas. Coral Reefs. 1997;**16**:225-230

[8] Reaka-Kudla ML. The global biodiversity of coral reefs: A comparison with rain forests. In: Reaka-Kudla ML, Wilson DE, Wilson EO, editors. Biodiversity II: Understanding and Protecting our Biological Resources. Washington DC: Joseph Henry Press; 1997. pp. 83-108

[9] Odum HT, Odum EP. Trophic structure and productivity of a windward coral reef community on Eniwetok atoll. Ecological Monographs. 1955;**25**:291-320

[10] Hatcher BG. Coral reef primary productivity. A hierarchy of pattern and process. Trends in Ecology & Evolution. 1990;**5**:149-155

[11] Atkinson MJ, Falter JL. In: Black KD, Shimmield GB, editors. Biogeochemistry of Marine Systems. Oxford: Blackwell Publishing; 2003. pp. 40-64

[12] McAllister DE. What is the status of the world's coral reef fishes? Sea Wind. 1991;**5**:14-18

[13] Knowlton N, Brainard RE, Fisher R, Moews M, Plaisance L, Caley MJ. Coral reef biodiversity. In: McIntyre A, editor. Life in the World's Oceans: Diversity Distribution and Abundance. Scotland, UK: John Wiley & Sons; 2010. pp. 65-74

[14] Wild C, Huettel M, Klueter A, Kremb SG, Rasheed MY, Jørgensen BB. Coral mucus functions as an energy carrier and particle trap in the reef ecosystem. Nature. 2004;**428**:66-70

[15] Haas AF, Naumann MS, Struck U, Mayr C, el-Zibdah M, Wild C. Organic matter release by coral reef associated benthic algae in the northern Red Sea. Journal of Experimental Marine Biology and Ecology. 2010;**389**:53-60

[16] Azam F, Fenchel T, Field JG, Gray JS, Meyer-Rei LA, Thingstad F. The ecological role of water-column microbes in the sea. Marine Ecology Progress Series. 1983;**10**:257-263

[17] De Goeij JM, Van Oevelen D, Vermeij MJ, Osinga R, Middelburg JJ, De Goeij AF, et al. Surviving in a marine desert: The sponge loop retains resources within coral reefs. Science. 2013;**342**(6154):108-110

[18] Buddemeier RW, Baker AC, Fautin DG, Jacobs JR. The adaptive hypothesis of bleaching. In: Rosenberg E, Loya Y, editors. Coral

Health and Disease. Berlin: Springer; 2004. pp. 427-444

[19] Bellwood DR, Hughes TP, Hoey AS. Sleeping functional group drives coral-reef recovery. Current Biology. 2006; **16**(24):2434-2439

[20] Connell JH. Diversity in tropical rain forests and coral reefs. Science. 1978;**199**(4335):1302-1310

[21] Lesser MP. Coral reef bleaching and global climate change: Can corals survive the next century? Proceedings of the National Academy of Sciences of the United States of America. 2007;**104**: 5259-5260

[22] Lough JM, Anderson KD, Hughes TP. Increasing thermal stress for tropical coral reefs: 1871–2017. Scientific Reports. 2018;**8**(1):6079

[23] Wilkinson C. Status of Coral Reefs of the World: 2008. Australia, Townsville: Global Coral Reef Monitoring Network and Reef and Rainforest Research Center; 2008

[24] Selig ER, Casey KS, Bruno JF. Temperature-driven coral decline: The role of marine protected areas. Global Change Biology. 2012;**18**(5):1561-1570

[25] Hughes TP, Barnes ML, Bellwood DR, Cinner JE, Cumming GS, Jackson JB, et al. Coral reefs in the Anthropocene. Nature. 2017;**546** (7656):82

[26] Hughes TP, Kerry JT, Álvarez-Noriega M, Álvarez-Romero JG, Anderson KD, Baird AH, et al. Global warming and recurrent mass bleaching of corals. Nature. 2017;**543**(7645):373

[27] Hughes TP, Anderson KD, Connolly SR, Heron SF, Kerry JT, Lough JM, et al. Spatial and temporal patterns of mass bleaching of corals in the Anthropocene. Science. 2018; **359**(6371):80-83

[28] Graham NA, Wilson SK, Jennings S, Polunin NV, Robinson JA, Bijoux JP, et al. Lag effects in the impacts of mass coral bleaching on coral reef fish, fisheries, and ecosystems. Conservation Biology. 2007;**21**(5):1291-1300

[29] Cardinale BJ, Duffy JE, Gonzalez A, Hooper DU, Perrings C, Venail P, et al. Biodiversity loss and its impact on humanity. Nature. 2012;**486**(7401):59

[30] Hughes TP, Tanner JE. Recruitment failure, life histories, and long-term decline of Caribbean corals. Ecology. 2000;**81**:2250-2263

[31] Vermeij MJA. Early life-history dynamics of Caribbean coral species on artificial substratum: The importance of competition, growth and variation in life-history strategy. Coral Reefs. 2006; **25**:5-71

[32] Darling ES, Alvarez-Filip L, Oliver TA, McClanahan TR, Côté IM. Evaluating life-history strategies of reef corals from species traits. Ecology Letters. 2012;**15**:1378-1386

[33] Darling ES, McClanahan TR, Côté IM. Life histories predict coral community disassembly under multiple stressors. Global Change Biology. 2013; **19**:1930-1940

[34] Kayal M, Vercelloni J, Wand MP, Adjeroud M. Searching for the best bet in life-strategy: A quantitative approach to individual performance and population dynamics in reef-building corals. Ecological Complexity. 2015;**23**: 73-84

[35] García-Arredondo A, Rojas A, Iglesias-Prieto R, Zepeda-Rodriguez A, Palma-Tirado L. Structure of nematocysts isolated from the fire corals *Millepora alcicornis* and *Millepora complanata* (Cnidaria: Hydrozoa). Journal of Venomous Animals and Toxins Including Tropical Diseases. 2012;**18**(1):109-115

[36] Rojas-Molina A, García-Arredondo A, Ibarra-Alvarado C, Bah M. *Millepora* ("fire corals") species: Toxinological studies until 2011. Advances in Environmental Research. 2012;**26**: 133-148

[37] Nagelkerken I, Nagelkerken WP. Loss of coral cover and biodiversity on shallow *Acropora* and *Millepora* reefs after 31 years on Curaçao, Netherlands Antilles. Bulletin of Marine Science. 2004;**74**:213-223

[38] Coni EO, Ferreira CM, de Moura RL, Meirelles PM, Kaufman L, Francini-Filho RB. An evaluation of the use of branching fire-corals (*Millepora* spp.) as refuge by reef fish in the Abrolhos Bank, eastern Brazil. Environmental Biology of Fishes. 2013; **96**(1):45-55

[39] Kayal E, Roure B, Philippe H, Collins AG, Lavrov DV. Cnidarian phylogenetic relationships as revealed by mitogenomics. BMC Evolutionary Biology. 2013;**13**(1):5

[40] Bouillon J. Essai de classification des hydropolypes-hydroméduses (Hydrozoa-Cnidaria). Indo-Malayan Zoology. 1985;**1**:29-243

[41] Schuchert P. The Hydrozoa Directory [Internet]. 2005. Available from: http://www.ville-ge.ch/musinfo/mhng/hydrozoa/classification.htm. [Accessed 2019-02-09]

[42] Bouillon J, Gravili C, Pagès F, Gili J-M, Boero F. An introduction to Hydrozoa. Mémoires du Muséum National d'Histoire Naturelle Paris. 2006;**194**:1-591

[43] Linnaeus C. Systema Naturae. Vol. I. Laurentii Salvii: Holmiae; 1758

[44] Duchassaing P, Michelotti J. Mémoire sur les Coralliaires des Antilles. Mémoire sur les coralliaires des Antilles 2. 1860;**19**:279-365

[45] Hickson SJ. Notes on the collection of specimens of the genus *Millepora*, obtained by Mr. Stanley Gardiner at Funafuti and Rotuma. In: Proceedings of the Zoological Society of London. Vol. 66. 1898. pp. 828-833

[46] Boschma H. The species problem in *Millepora*. Zoologische Verhandelingen. 1948;**1**:1-115

[47] Boschma H. On a new species of *Millepora* from Mauritius, with notes on the specific characters of *Millepora* exaesa. In: Proceedings of the Koninklijke Nedderlandse Akademie van Wetenschappen. Vol. 69. 1966. pp. 409-419

[48] Heckenhauer J, Schweinsberg M, Elbrecht V, John U, Tollrian R, Lampert KP. Isolation, characterization and cross amplification of eleven novel microsatellite loci for the hydrozoan coral *Millepora*. Conservation Genetics Resources. 2015;7:215-217

[49] Tepper CS, Gaynor SC. Ribosomal internal transcribed spacer (ITS) DNA variation in *Millepora*. Journal of Marine Science Research and Development. 2015;**6**:1

[50] Dubé CE, Planes S, Zhou Y, Berteaux-Lecellier V, Boissin E. Genetic diversity and differentiation in reef-building *Millepora* species, as revealed by cross-species amplification of fifteen novel microsatellite loci. PeerJ. 2017;**5**: e2936

[51] Lewis JB. Biology and ecology of the hydrocoral *Millepora* on coral reefs. Marine Biology. 2006;**50**:1-53

[52] Boschma H. Milleporina and Stylasterina. In: Moore RC, editor. Treatise on Invertebrate Paleontology. Part F, Coelenterata. London: Geological Society of America and University of Kansas Press; 1956

[53] Vaughan TW. Fossil corals from Central America, Cuba and Puerto Rico,

with an account of the American tertiary, Pleistocene, and recent coral reefs. United States National Museum Bulletin. 1919;**103**:189-524

[54] Rehfeld U, Ernst G. Hydrozoan build-ups of *Millepora irregularis* sp. nov. and fungiid coral meadows of Cunnolites Alloiteau (Anthozoa) - Palaeoecological and palaeoceanographical implications for the upper cretaceous of North Cantabria (northern Spain). Facies. 1998;**39**: 125-138

[55] Florez P, Zapata-Ramirez P, Klaus JS. Early Miocene shallow-water corals from La Guajira, Colombia: Part II, Mussidae-Siderastreidae and Milleporidae. Journal of Paleontology. 2019;**93**:416-436

[56] Vyverberg K, Dechnik B, Dutton A, Webster JM, Zwartz D, Portell RW. Episodic reef growth in the granitic Seychelles during the last interglacial: Implications for polar ice sheet dynamics. Marine Geology. 2018;**399**: 170-187

[57] Arrigoni R, Maggioni D, Montano S, Hoeksema BW, Seveso D, Shlesinger T, et al. An integrated morpho-molecular approach to delineate species boundaries of *Millepora* from the Red Sea. Coral Reefs. 2018;**37**(4):967-984

[58] Dubé CE, Boissin E, Maynard JA, Planes S. Fire coral clones demonstrate phenotypic plasticity among reef habitats. Molecular Ecology. 2017;**26**: 3860-3869

[59] Dubé CE, Boissin E, Planes S. Overgrowth of living scleractinian corals by the hydrocoral *Millepora platyphylla* in Moorea, French Polynesia. Marine Biodiversity. 2016; **46**(2):329-330

[60] Wegener C, Martin B, Didden C, Edmunds PJ. Overgrowth of Caribbean octocorals by milleporid hydrocorals. Invertebrate Biology. 2018;**137**(1):29-37

[61] WoRMS 2019. Millepora Linnaeus [Internet]. 1758. Available from: http://www.marinespecies.org/aphia.php?p=taxdetails&id=205902 [Accessed 2019-02-09]

[62] Duchassaing P, Michelotti J, et al. Supplément au Mémoire sur les coralliaires des Antilles 2. 1864;**23**:1-256

[63] Hickson SJ. Report on the specimens of the genus *Millepora* collected by Dr. Willey. Journal of Zoology. 1902;**119**: 661-672

[64] Duchassaing P. Animaux radiaires des Antilles. Paris: Typographie Plon Fréres; 1850. 48 p

[65] Razak TB, Hoeksema BW. The hydrocoral genus *Millepora* (Hydrozoa: Capitata: Milleporidae) in Indonesia. Zoologische Verhandelingen Leiden. 2003;**345**:313-336

[66] Moschencko AV. Anatomy and morphology of skeleton and soft tissues of *Millepora* spp. (Hydrozoa, Athecata, Milleporidae). Zoologicheskiĭ Zhurnal. 1993;**72**:5-14

[67] Ruiz-Ramos DV, Weil E, Schizas NV. Morphological and genetic evaluation of the hydrocoral *Millepora* species complex in the Caribbean. Zoological Studies. 2014;**53**:15

[68] De Souza JN, Nunes FL, Zilberberg C, Sanchez JA, Migotto AE, Hoeksema BW, et al. Contrasting patterns of connectivity among endemic and widespread fire coral species (*Millepora* spp.) in the tropical Southwestern Atlantic. Coral Reefs. 2017;**36**(3):701-716

[69] Bourmaud CA-F, Leung JKL, Bollard S, Gravier-Bonnet N. Mass spawning events, seasonality and reproductive features in Milleporids (Cnidaria, Hydrozoa) from Reunion Island. Marine Ecology. 2013;**34**:14-24

[70] Shlesinger T, Loya Y. Mass medusae release and temporal reproductive

segregation among the three Red Sea fire coral species. Bulletin of the Ecological Society of America. 2019; **100**(2):1-3

[71] Hebert PDN, Ratnasingham S, de Waard JR. Barcoding animal life: Cytochrome c oxidase subunit 1 divergences among closely related species. Proceedings of the Royal Society of London B: Biological Sciences. 2003;**270**:S96-S99

[72] Hoareau TB, Boissin E. Design of phylum-specific hybrid primers for DNA barcoding: Addressing the need for efficient COI amplification in the Echinodermata. Molecular Ecology Resources. 2010;**10**:960-967

[73] Bucklin A, Steinke D, Blanco-Bercial L. DNA barcoding of marine Metazoa. In: Carlson CA, Giovannoni SJ, editors. Annual Review of Marine Science. Vol. 3. Palo Alto: Annual Reviews; 2011. pp. 471-508

[74] Geller J, Meyer C, Parker M, Hawk H. Redesign of PCR primers for mitochondrial cytochrome c oxidase subunit I for marine invertebrates and application in all-taxa biotic surveys. Molecular Ecology Resources. 2013;**13**: 851-861

[75] Amaral FD, Steiner AQ, Broadhurst MK, Cairns SD. An overview of the shallow-water calcified hydroids from Brazil (Hydrozoa: Cnidaria), including the description of a new species. Zootaxa. 1930;**2008**:56-68

[76] Hoeksema BW, Nunes FLD, Lindner A, De Souza JN. *Millepora alcicornis* (Hydrozoa: Capitata) at Ascension Island: Confirmed identity based on morphological and molecular analyses. Journal of the Marine Biological Association of the United Kingdom. 2017;**97**:709-712

[77] Boissin E, Pogoreutz C, Pey A, Gravier-Bonnet N, Planes S. *Millepora platyphylla* (Cnidaria, Hydrozoa) range extended back to the eastern Pacific, thanks to a new record from Clipperton atoll. Zootaxa. 2019;**4668**:599-600

[78] Glynn PW, de Weerdt WH. Elimination of 2 reef-building hydrocorals following the 1982-83 El-Nino warming event. Science. 1991;**253**: 69-71

[79] Clemente S, Rodríguez A, Brito A, Ramos A, Monterroso Ó, Hernández JC. On the occurrence of the hydrocoral *Millepora* (Hydrozoa: Milleporidae) in the subtropical eastern Atlantic (Canary Islands): Is the colonization related to climatic events? Coral Reefs. 2011;**30**(1): 237-240

[80] Jokiel PL. Rafting of reef corals and other organisms at Kwajalein atoll. Marine Biology. 1989;**101**:483-493

[81] López C, Clemente S, Almeida C, Brito A, Hernández M. A genetic approach to the origin of *Millepora* sp. in the eastern Atlantic. Coral Reefs. 2015; **34**(2):631-638

[82] Lewis JB. The ecology of *Millepora*. Coral Reefs. 1989;**8**:99-107

[83] Cairns SD. Worldwide distribution of the Stylasteridae (Cnidaria: Hydrozoa). Scientia Marina. 1992;**56**: 125-130

[84] Dustan P, Halas JC. Changes in the reef-coral community of Carysfort reef, Key largo, Florida: 1974 to 1982. Coral Reefs. 1987;**6**(2):91-106

[85] Loya Y, Sakai K, Yamazato K, Nakano Y, Sambali H, van Woesik R. Coral bleaching: The winners and the losers. Ecology Letters. 2001;**4**(2): 122-131

[86] Arias-González JE, Legendre P, Rodríguez-Zaragoza FA. Scaling up beta diversity on Caribbean coral reefs.

Journal of Experimental Marine Biology and Ecology. 2008;**366**(1–2):28-36

[87] Loya Y. Recolonization of Red Sea corals affected by natural catastrophes and man-made perturbations. Ecology. 1976;**57**(2):278-289

[88] Richard C, Benzoni F, Dutrieux E, Chaîneau CH. Marine Sensitivity Mapping of the Southern Coast of Yemen: From Balhaf to Mukallah, Creocean. Montpellier: UMR 227 Coreus 2, Labex CORAIL, Institut de Recherche pour le Développement; 2011. 101 p

[89] Andréfouët S, Benzoni F, Payri C. A monospecific *Millepora* reef in Marquesas Island, French Polynesia. Coral Reefs. 2014;**33**:463

[90] Rioja-Nieto R, Álvarez-Filip L. Coral reef systems of the Mexican Caribbean: Status, recent trends and conservation. Marine Pollution Bulletin. 2019;**140**:616-625

[91] Dubé CE, Mercière A, Vermeij MJ, Planes S. Population structure of the hydrocoral *Millepora platyphylla* in habitats experiencing different flow regimes in Moorea, French Polynesia. PLoS One. 2017;**12**:e0173513

[92] Strömgren T. Skeleton growth of the hydrocoral *Millepora complanata* Lamarck in relation to light. Limnology and Oceanography. 1976;**21**(1):156-160

[93] Meroz-Fine E, Brickner I, Loya Y, Ilan M. The hydrozoan coral *Millepora dichotoma*: Speciation or phenotypic plasticity? Marine Biology. 2003;**143**: 1175-1183

[94] Monismith SG. Hydrodynamics of coral reefs. Annual Review of Fluid Mechanics. 2007;**39**:37-55

[95] Ferrario F, Beck MW, Storlazzi CD, Micheli F, Shepard CC, Airoldi L. The effectiveness of coral reefs for coastal hazard risk reduction and adaptation. Nature Communications. 2014;**13**:5

[96] Lewis JB. Spatial distributions of the calcareous hydrozoans *Millepora complanata* and *Millepora squarrosa* on coral reefs. Bulletin of Marine Science. 1996;**59**(1):188-195

[97] Faure G. Recherches sur le peuplement de scléractiniaires des récifs coralliens de l'archipel des Mascareignes (Ocean Indien occidental) [thesis]. Marseille: Université Aix-Marseille II; 1982

[98] Garcia TM, Matthews-Cascon H, Franklin-Junior W. *Millepora alcicornis* (Cnidaria: Hydrozoa) as substrate for benthic fauna. Brazilian Journal of Oceanography. 2009;**57**:153-155

[99] Stella JS, Jones GP, Pratchett MS. Variation in the structure of epifaunal invertebrate assemblages among coral hosts. Coral Reefs. 2010;**29**(4):957-973

[100] Glynn PW, Enochs IC. Invertebrates and their roles in coral reef ecosystems. In: Coral Reefs: An Ecosystem in Transition. Dordrecht: Springer; 2011. pp. 273-325

[101] Pereira PHC, Leal ICS, de Araújo ME, Souza AT. Feeding association between reef fishes and the fire coral *Millepora* spp. (Cnidaria: Hydrozoa). Marine Biodiversity Records. 2012;**5**:e42

[102] Leal ICS, Pereira PHC, De Araujo ME. Coral reef fish association and behaviour on the fire coral *Millepora* spp. in north-East Brazil. Journal of the Marine Biological Association of the United Kingdom. 2013;**93**:1703-1711

[103] Vermeij MJ, DeBey H, Grimsditch G, Brown J, Obura D, DeLeon R, et al. Negative effects of gardening damselfish *Stegastes planifrons* on coral health depend on predator

abundance. Marine Ecology Progress Series. 2015;**528**:289-296

[104] Hoeksema BW, Ten Hove HA. Attack on a Christmas tree worm by a Caribbean sharpnose pufferfish at St. Eustatius, Dutch Caribbean. Bulletin of Marine Science. 2017;**93**(4):1023-1024

[105] Perry O, Sapir Y, Perry G, Ten Hove H, Fine M. Substrate selection of Christmas tree worms (*Spirobranchus* spp.) in the Gulf of Eilat, Red Sea. Journal of the Marine Biological Association of the United Kingdom. 2018;**98**(4):791-799

[106] Kayal M, Kayal E. Colonies of the fire coral *Millepora platyphylla* constitute scleractinian survival oases during *Acanthaster* outbreaks in French Polynesia. Marine Biodiversity. 2017; **47**(1):255-258

[107] Bertelsen E, Ussing H. Marine tropical animals carried to the Copenhagen Sydhavn on a ship from the Bermudas. Vidensk Medd Dansk Naturhist Foren Kobenhavn. 1936;**100**: 237-245

[108] Wahle CM. Detection, pursuit, and overgrowth of tropical gorgonians by milleporid hydrocorals: Perseus and Medusa revisited. Science. 1980;**209**: 689-691

[109] Glynn PW. Feeding ecology of selected coral-reef macroconsumers: Patterns and effects on coral reef community structure. In: Dubinsky Z, editor. Ecosystems of the World: Coral Reefs. New York: Elsevier; 1990. pp. 365-400

[110] LaJeunesse TC, Parkinson JE, Gabrielson PW, Jeong HJ, Reimer JD, Voolstra CR, et al. Systematic revision of Symbiodiniaceae highlights the antiquity and diversity of coral endosymbionts. Current Biology. 2018; **28**(16):2570-2580

[111] Hoegh-Guldberg O. Climate change, coral bleaching and the future of the world's coral reefs. Marine and Freshwater Research. 1999;**50**(8): 839-866

[112] Weis V, Reynolds W, de Boer M, Krupp D. Host-symbiont specificity during onset of symbiosis between the dinoflagellates *Symbiodinium* spp. and planula larvae of the scleractinian coral *Fungia scutaria*. Coral Reefs. 2001;**20**: 301-308

[113] Yellowlees D, Rees TAV, Leggat W. Metabolic interactions between algal symbionts and invertebrate hosts. Plant, Cell and Environment. 2008;**31**:679-694

[114] Davy SK, Allemand D, Weis VM. Cell biology of cnidarian-Dinoflagellate symbiosis. Microbiology and Molecular Biology Reviews. 2012;**76**:229-261

[115] Mieog JC, Olsen JL, Berkelmans R, Bleuler-Martinez SA, Willis BL, van Oppen MJH. The roles and interactions of symbiont, host and environment in defining coral fitness. PLoS One. 2009; **4**:e6364

[116] Parkinson JE, Banaszak AT, Altman NS, LaJeunesse TC, Baums IB. Intraspecific diversity among partners drives functional variation in coral symbioses. Scientific Reports. 2015;**5**: 1-12

[117] Lewis JB. Heterotrophy in corals: Zooplankton predation by the hydrocoral *Millepora complanata*. Marine Ecology Progress Series. 1993; **90**:251

[118] Banaszak AT, Santos MGB, LaJeunesse TC, Lesser MP. The distribution of mycosporine-like amino acids (MAAs) and the phylogenetic identity of symbiotic dinoflagellates in cnidarian hosts from the Mexican Caribbean. Journal of Experimental Marine Biology and Ecology. 2006;**332**: 31-146

[119] van Oppen MJ, Baker AC, Coffroth MA, Willis BL. Bleaching resistance and the role of algal endosymbionts. In: Coral Bleaching. Berlin: Heidelberg; 2009. pp. 83-102

[120] Baker AC. Flexibility and specificity in coral-algal symbiosis: Diversity, ecology, and biogeography of *Symbiodinium*. Annual Review of Ecology, Evolution, and Systematics. 2003;**34**:661-689

[121] Samayoa A, Reyes S, Karim YB, Roge-Jones L, Rueth M, Tepper C. Patterns of Millepore-*Symbiodinium* associations at two Caribbean locations: San Salvador, the Bahamas and south water Caye, Belize. In: Proceedings of the First Joint Natural History and Geology Symposium. 2017

[122] Rodríguez L, López C, Casado-Amezua P, Ruiz-Ramos DV, Martínez B, Banaszak A, et al. Genetic relationships of the hydrocoral *Millepora alcicornis* and its symbionts within and between locations across the Atlantic. Coral Reefs. 2019;**38**(2):255-268

[123] Lesser MP. Coral bleaching: Causes and mechanisms. In: Dubinsky Z, Stambler N, editors. Coral Reefs: An Ecosystem in Transition. Dordrecht: Springer; 2011. pp. 405-419

[124] Bonesso JL, Leggat W, Ainsworth TD. Exposure to elevated sea-surface temperatures below the bleaching threshold impairs coral recovery and regeneration following injury. PeerJ. 2017;**5**:e3719

[125] Neal BP, Khen A, Treibitz T, Beijbom O, O'Connor G, Coffroth MA, et al. Caribbean massive corals not recovering from repeated thermal stress events during 2005-2013. Ecology and Evolution. 2017;7(5):1339-1353

[126] Oliver JK, Berkelmans R, Eakin CM. Coral bleaching in space and time. In: Coral Bleaching. Cham: Springer; 2018. pp. 27-49

[127] Baker AC, Glynn PW, Riegl B. Climate change and coral reef bleaching: An ecological assessment of long-term impacts, recovery trends and future outlook. Estuarine, Coastal and Shelf Science. 2008;**80**(4):435-471

[128] Grottoli AG, Warner ME, Levas SJ, Aschaffenburg MD, Schoepf V, McGinley M, et al. The cumulative impact of annual coral bleaching can turn some coral species winners into losers. Global Change Biology. 2014; **20**(12):3823-3833

[129] Okazaki RR, Towle EK, Van Hooidonk R, Mor C, Winter RN, Piggot AM, et al. Species-specific responses to climate change and community composition determine future calcification rates of Florida keys reefs. Global Change Biology. 2016; **23**(3):1023-1035

[130] Marshall PA, Baird AH. Bleaching of corals on the great barrier reef: Differential susceptibilities among taxa. Coral Reefs. 2000;**19**:155-163

[131] Rowan R, Knowlton N, Baker A, Jara J. Landscape ecology of algal symbionts creates variation in episodes of coral bleaching. Nature. 1997; **388**(6639):265

[132] Guest JR, Baird AH, Maynard JA, Muttaqin E, Edwards AJ, Campbell SJ, et al. Contrasting patterns of coral bleaching susceptibility in 2010 suggest an adaptive response to thermal stress. PLoS One. 2012;**7**(3):e33353

[133] Dias TL, Gondim AI. Bleaching in scleractinians, hydrocorals, and octocorals during thermal stress in a northeastern Brazilian reef. Marine Biodiversity. 2016;**46**(1):303-307

[134] Glynn PW. Coral reef bleaching: Ecological perspectives. Coral Reefs. 1993;**12**(1):1-7

[135] Glynn PW, Feingold JS. Hydrocoral species not extinct. Science. 1992; **257**(5078):1845-1847

[136] Cook CB, Logan A, Ward J, Luckhurst B, Berg CJ. Elevated temperatures and bleaching on a high latitude coral reef: The 1988 Bermuda event. Coral Reefs. 1990;**9**(1):45-49

[137] Williams EH Jr, Bunkley-Williams L, et al. Atoll Research Bulletin. 1990; **335**:1-71

[138] Wagner DE, Kramer P, van Woesik R. Species composition, habitat, and water quality influence coral bleaching in southern Florida. Marine Ecology Progress Series. 2010;**408**:65-78

[139] Olguín-López N, Hérnandez-Elizárraga VH, Hernández-Matehuala R, Cruz-Hernández A, Guevara-González R, Caballero-Pérez J, et al. Impact of El Niño-southern oscillation 2015-2016 on the soluble proteomic profile and cytolytic activity of *Millepora alcicornis* ("fire coral") from the Mexican Caribbean. PeerJ. 2019;**7**: e6593

[140] van Woesik R, Sakai K, Ganase A, Loya Y. Revisiting the winners and the losers a decade after coral bleaching. Marine Ecology Progress Series. 2011; **434**:67-76

[141] Morri C, Bianchi CN, Aliani S. Coral reefs at Gangehi (north Ari atoll, Maldive Islands). Publications du Service Géologique du Luxembourg. 1995;**29**:3-12

[142] Coleman FC, Dennis GD, Jaap WC, Schmahl GP, Koenig CC, Reed SA, et al. Final Report to the National Oceanic and Atmospheric Administration Coral Reef Conservation Program. Part 1, Status and Trends in Habitat Characterization of the Florida Middle Grounds; 2004

[143] Wells JW, Davies PS. Preliminary list of stony corals from Addu atoll. Atoll Research Bulletin. 1966;**116**:43-55

[144] Coleman N. Marine Life of the Maldives. Sea Challengers. Appollo Bay: Atoll Editions; 2000

[145] McClanahan TR. Bleaching damage and recovery potential of Maldivian coral reefs. Marine Pollution Bulletin. 2000;**40**(7):587-597

[146] Bianchi CN, Morri C, Pichon M, Benzoni F, Colantoni P, Baldelli G, et al. Dynamics and pattern of coral recolonization following the 1998 bleaching event in the reefs of the Maldives. In: Proceedings of the 10th International Coral Reef Symposium, Okinawa, Japan. Vol. 1. 2006. pp. 30-37

[147] Bianchi CN, Morri C, Colantoni P, Sandrini M. Italian research in the Maldives. Reef Encounter. 2009;**37**:17

[148] Gravier-Bonnet N, Bourmaud CA-F. Hydroids (Cnidaria, Hydrozoa) of Baa atoll (Indian Ocean, Maldives archipelago). In: Biodiversity, Resources, and Conservation of Baa Atoll (Republic of Maldives): A Unesco Man and Biosphere Reserve. 2012

[149] Wilkinson C, Linden O, Cesar H, Hodgson G, Rubens J, Strong AE. Ecological and socioeconomic impacts of 1998 coral mortality in the Indian Ocean: An ENSO impact and a warning of future change? Ambio. 1999;**28**(2): 188-196

[150] Gleason MG. Effects of disturbance on coral communities: Bleaching in Moorea, French Polynesia. Coral Reefs. 1993;**12**(3–4):193-201

[151] Banaszak AT, Ayala-Schiaffino BN, Rodríguez-Román A, Enríquez S, Iglesias-Prieto R. Response of *Millepora alcicornis* (Milleporina: Milleporidae) to two bleaching events at Puerto Morelos reef, Mexican Caribbean. Revista de Biología Tropical. 2003;**51**(4):57-66

[152] Zvuloni A, Armoza-Zvuloni R, Shaked Y. Multifocal bleaching of *Millepora dichotoma* in the Gulf of Aqaba (Eilat), Red Sea. Marine Ecology Progress Series. 2011;**441**:25-32

[153] Paramasivam N, Ben-Dov E, Arotsker L, Kramarsky-Winter E,

Zvuloni A, Loya Y, et al. Bacterial consortium of *Millepora dichotoma* exhibiting unusual multifocal lesion event in the Gulf of Eilat, Red Sea. Microbial Ecology. 2013;**65**(1):50-59

[154] Bay RA, Palumbi SR. Rapid acclimation ability mediated by transcriptome changes in reef-building corals. Genome Biology and Evolution. 2015;7(6):1602-1612

[155] Pinzón JH, Kamel B, Burge CA, Harvell CD, Medina M, Weil E, et al. Whole transcriptome analysis reveals changes in expression of immune-related genes during and after bleaching in a reef-building coral. Royal Society Open Science. 2015;**2**(4):140214

[156] Seneca FO, Palumbi SR. The role of transcriptome resilience in resistance of corals to bleaching. Molecular Ecology. 2015;**24**(7):1467-1484

[157] Ruiz-Jones LJ, Palumbi SR. Tidal heat pulses on a reef trigger a fine-tuned transcriptional response in corals to maintain homeostasis. Science Advances. 2017;**3**(3):e1601298

[158] Huang Y, Yuan J, Zhang Y, Peng H, Liu L. Molecular cloning and characterization of calmodulin-like protein CaLP from the Scleractinian coral *Galaxea astreata*. Cell Stress & Chaperones. 2018;**23**(6):1329-1335

[159] Mayfield AB, Chen Y-J, Lu C-Y, Chen C-S. The proteomic response of the reef coral *Pocillopora acuta* to experimentally elevated temperatures. PLoS One. 2018;**13**(1):e0192001

[160] Brown D, Edmunds PJ. Differences in the responses of three scleractinians and the hydrocoral *Millepora platyphylla* to ocean acidification. Marine Biology. 2016;**163**(3):62

[161] Marangoni LFB, Calderon EN, Marques JA, Duarte GAS, Pereira CM, Castro CB, et al. Effects of CO_2-driven acidification of seawater on the calcification process in the calcareous hydrozoan *Millepora alcicornis* (Linnaeus, 1758). Coral Reefs. 2017;**36**: 1133-1141

[162] Kroeker KJ, Kordas RL, Crim R, Hendriks IE, Ramajo L, Singh GS, et al. Impacts of ocean acidification on marine organisms: Quantifying sensitivities and interaction with warming. Global Change Biology. 2013; **19**(6):1884-1896

[163] Chan NCS, Connolly SR. Sensitivity of coral calcification to ocean acidification: A meta-analysis. Global Change Biology. 2013;**19**(1):282-290

[164] Kaniewska P, Campbell PR, Kline DI, Rodriguez-Lanetty M, Miller DJ, Dove S, et al. Major cellular and physiological impacts of ocean acidification on a reef building coral. PLoS One. 2012;**7**:e34659

[165] Soriano-Santiago OS, Liñán-Cabello MA, Delgadillo-Nuño MA, Ortega-Ortiz C, Cuevas-Venegas S. Physiological responses to oxidative stress associated with pH variations in host tissue and zooxanthellae of hermatypic coral *Pocillopora capitata*. Marine and Freshwater Behaviour and Physiology. 2013;**46**(5):275-286

[166] Luz DC, Zebral YD, Klein RD, Marques JA, de Barros Marangoni LF, Pereira CM, et al. Oxidative stress in the hydrocoral *Millepora alcicornis* exposed to CO_2-driven seawater acidification. Coral Reefs. 2018;**37**(2):571-579

[167] Downs CA, Fauth JE, Halas JC, Dustan P, Bemiss J, Woodley CM. Oxidative stress and seasonal coral bleaching. Free Radical Biology and Medicine. 2002;**33**:533-543

[168] Todd PA. Morphological plasticity in scleractinian corals. Biological Reviews of the Cambridge Philosophical Society. 2008;**83**:315-337

[169] Denny MW. Biology and the Mechanics of the Wave-Swept Environment. New Jersey: Princeton University Press; 2014

[170] Swierts T, Vermeij MJA. Competitive interactions between corals and turf algae depend on coral colony form. PeerJ. 2016;**4**:e1984

[171] Baird AH, Hughes TP. Competitive dominance by tabular corals: An experimental analysis of recruitment and survival of understorey assemblages. Journal of Experimental Marine Biology and Ecology. 2000;**251**: 117-132

[172] Madin JS. Mechanical limitations of reef corals during hydrodynamic disturbances. Coral Reefs. 2005;**24**: 630-635

[173] Madin JS, Baird AH, Dornelas M, Connolly SR. Mechanical vulnerability explains size-dependent mortality of reef corals. Ecology Letters. 2014;**17**: 1008-1015

[174] Ayre DJ, Willis BL. Population structure in the coral *Pavona cactus*: Clonal genotypes show little phenotypic plasticity. Marine Biology. 1988;**99**: 495-505

[175] West JM, Harvell CD, Walls AM. Morphological plasticity in a gorgonian coral (*Briareum asbestinum*) over a depth cline. Marine Ecology Progress Series. 1993;**94**:61-69

[176] Bruno JF, Edmunds PJ. Clonal variation for phenotypic plasticity in the coral *Madracis mirabilis*. Ecology. 1997; **78**:2177-2190

[177] Hill MS, Hill AL. Morphological plasticity in the tropical sponge *Anthosigmella varians*: Responses to predators and wave energy. The Biological Bulletin. 2002;**202**:86-95

[178] Pineda-Krch M, Lehtilä K. Costs and benefits of genetic heterogeneity within organisms. Journal of Evolutionary Biology. 2004;**17**:1167-1177

[179] Agrawal AA. Phenotypic plasticity in the interactions and evolution of species. Science. 2001;**294**:321-326

[180] Beaman JE, White CR, Seebacher F. Evolution of plasticity: Mechanistic link between development and reversible acclimation. Trends in Ecology & Evolution. 2016;**31**:237-249

[181] Edmunds PJ. The role of colony morphology and substratum inclination in the success of *Millepora alcicornis* on shallow coral reefs. Coral Reefs. 1999;**18**: 133-140

[182] Jackson JBC. Morphological strategies of sessile animals. In: Larwood G, Rosen BR, editors. Biology and Systematics of Colonial Organisms. London: Academic Press; 1979. pp. 499-555

[183] Highsmith RC. Reproduction by fragmentation in corals. Marine Ecology Progress Series. 1982;**7**:207-226

[184] Jackson JBC. Modes of dispersal of clonal benthic invertebrates: Consequences for 4 species' distributions and genetic structure of local populations. Bulletin of Marine Science. 1986;**39**:588-605

[185] Sherman CDH, Ayre DJ, Miller KJ. Asexual reproduction does not produce clonal populations of the brooding coral *Pocillopora damicornis* on the great barrier reef, Australia. Coral Reefs. 2006;**25**:7-18

[186] Foster NL, Baums IB, Sanchez JA, Paris CB, Chollett I, Agudelo CL, et al. Hurricane-driven patterns of clonality in an ecosystem engineer: The Caribbean coral *Montastraea annularis*. PLoS One. 2013;**8**(1):e53283

[187] Harrison PL. Sexual reproduction of scleractinian corals. In: Zvy D,

Stambler N, editors. Coral Reefs: An Ecosystem in Transition. Dordrecht: Springer Netherlands; 2011. pp. 59-85

[188] Gravier-Bonnet N, Bourmaud CA-F. Cloning by releasing specialized frustules in a successful epiphytic zooxanthellate halecid (Cnidaria, Hydrozoa, Haleciidae), with comments on stolonization and frustulation. Journal of Invertebrate Reproduction and Development. 2005;**48**:63-69

[189] Pearson EA, Murray SN. Patterns of reproduction, genetic diversity, and genetic differentiation in California populations of the geniculate coralline alga *Lithothrix aspergillum* (Rhodophyta). Journal of Phycology. 1997;**33**:753-763

[190] Sherman CDH, Ayre DJ. Fine-scale adaptation in a clonal sea anemone. Evolution. 2008;**62**:1373-1380

[191] Hoareau TB, Boissin E, Conand C. First record of asexual reproduction of *Holothuria hilla* in a fringing reef of Reunion Island. SPC Beche-de-Mer Information Bulletin. 2010;**27**:24

[192] Kahng SE, Benayahu Y, Lasker HR. Sexual reproduction in octocorals. Marine Ecology Progress Series. 2011; **443**:265-283

[193] Whalan S, Johnson MS, Harvey E, Battershill C. Mode of reproduction, recruitment, and genetic subdivision in the brooding sponge *Haliclona* sp. Marine Biology. 2005;**146**:425-433

[194] Miller KJ, Ayre DJ. The role of sexual and asexual reproduction in structuring high latitude populations of the reef coral *Pocillopora damicornis*. Heredity. 2004;**92**:557-568

[195] Gorospe KD, Karl SA. Genetic relatedness does not retain spatial pattern across multiple spatial scales: Dispersal and colonization in the coral *Pocillopora damicornis*. Molecular Ecology. 2013;**22**:3721-3736

[196] Baums IB, Devlin-Durante M, Laing BA, Feingold J, Smith T, Bruckner A, et al. Marginal coral populations: The densest known aggregation of *Pocillopora* in the Galápagos archipelago is of asexual origin. Frontiers in Marine Science. 2014;**1**:59

[197] Rice WR, Chippindale AK. Sexual recombination and the power of natural selection. Science. 2001;**294**:555-559

[198] Williams GC. Sex and Evolution. Princeton University Press: Princeton; 1975

[199] Brazeau DA, Lasker HR. Sexual reproduction and external brooding by the Caribbean gorgonian *Briareum asbestinum*. Marine Biology. 1990;**104**: 465-474

[200] Ereskovsky AV, Tokina DB. Asexual reproduction in homoscleromorph sponges (Porifera; Homoscleromorpha). Marine Biology. 2007;**151**:425-434

[201] Reitzel AM, Burton PM, Krone C, Finnerty JR. Comparison of developmental trajectories in the starlet sea anemone *Nematostella vectensis*: Embryogenesis, regeneration, and two forms of asexual fission. Invertebrate Biology. 2007;**126**:99-112

[202] Gutiérrez-Rodríguez C, Lasker HR. Reproductive biology, development, and planula behavior in the Caribbean gorgonian *Pseudopterogorgia elisabethae*. Invertebrate Biology. 2004;**123**:54-67

[203] Nakajima Y, Chuang PS, Ueda N, Mitarai S. First evidence of asexual recruitment of *Pocillopora acuta* in Okinawa Island using genotypic identification. PeerJ. 2018;**6**:e5915

[204] Oury N, Gélin P, Massé L, Magalon H. First study of asexual planulae in the coral *Pocillopora damicornis* type β SSH05c from the

southwestern Indian Ocean. Coral Reefs. 2019;**38**(3):499-503

[205] Nomura J. Sexual Reproduction of *Millepora intricata* and *Millepora tenella* (Hydrozoa : Milleporidae) [Thesis]. Okinawa: University of the Ryukyus; 1998

[206] Soong K, Cho LC. Synchronized release of medusae from three species of hydrozoan fire corals. Coral Reefs. 1998; **17**:145-154

[207] Boschma H. The ampullae of *Millepora*. In: Proceedings Koninklijke Nederlandse Akademie van Wetenschappen Amsterdam. Vol. 3. 1949. pp. 2-3

[208] Boschma H. Further notes on the ampullae of *Millepora*. Zoologische Mededelingen Leiden. 1950;**31**:49-61

[209] de Weerdt WH. Taxonomic characters in Caribbean *Millepora* species (Hydrozoa, Coelenterata). Bijdragen tot de Dierkunde. 1984;**54**: 243-262

[210] Quelch JJ. The Milleporidae. Nature. 1884;**30**:539

[211] Quelch JJ. Report of the reef-corals collected by H. M. S. challenger during the years 1873-76. Reports of the Challenger, Zoology. 1886;**6**:189-194

[212] Boschma H. Notes on the ampullae of two colonies of *Millepora*. In: Proceedings Koninklijke Nederlandse Akademie van Wetenschappen Amsterdam. Vol. 67. 1964. pp. 195-200

[213] Gili J-M, Hughes RG. The ecology of marine benthic hydroids. Oceanography and Marine Biology: An Annual Review. 1995;**33**:351-426

[214] Tsounis G, Rossi S, Aranguren M, Gili JM, Arntz W. Effects of spatial variability and colony size on the reproductive output and gonadal development cycle of the Mediterranean red coral (*Corallium rubrum L.*). Marine Biology. 2006;**148**:513-527

[215] Gaines SD, Gaylord B, Largier JL. Avoiding current oversights in marine reserve design. Ecological Applications. 2003;**13**:32-46

[216] Ayre DJ, Hughes TP. Climate change, genotypic diversity and gene flow in reef-building corals. Ecology Letters. 2004;**7**:273-278

[217] Steneck RS. Staying connected in a turbulent world. Science. 2006;**311**: 480-481

[218] Adjeroud M, Guérécheau A, Vidal-Dupiol J, Flot JF, Arnaud-Haond S, Bonhomme F. Genetic diversity, clonality and connectivity in the scleractinian coral *Pocillopora damicornis*: A multi-scale analysis in an insular, fragmented reef system. Marine Biology. 2014;**161**(3):531-541

[219] Combosch DJ, Vollmer SV. Population genetics of an ecosystem-defining reef coral *Pocillopora damicornis* in the tropical eastern Pacific. PLoS One. 2011;**6**:e21200

[220] Pinzón JH, Reyes-Bonilla H, Baums IB, Lajeunesse TC. Contrasting clonal structure among *Pocillopora* (Scleractinia) communities at two environmentally distinct sites in the Gulf of California. Coral Reefs. 2012;**3**: 765-777

[221] Dubé CE, Boissin E, Mercière A, Planes S. Dispersal limitations of early life stages and sibling aggregations in a broadcasting *Millepora* hydrocoral, as revealed by parentage analysis. bioRxiv. 2018;**1**:413963

[222] Hughes RN. Lessons in modularity: The evolutionary ecology of colonial invertebrates. Scientia Marina. 2005;**69**: 169-179

[223] Jackson JBC. Competition on marine hard substrata: The adaptive significance of solitary and colonial strategies. American Naturalist. 1977; **111**:743-767

[224] Jackson JBC. Distribution and ecology of clonal and aclonal benthic invertebrates. In: Jackson JBC, Buss LW, Cook RE, editors. Population Biology and Evolution of Clonal Organisms. New Haven: Yale University Press; 1985. pp. 297-355

[225] Hughes TP, Jackson JBC. Do corals lie about their age? Some demographic consequences of partial mortality, fission and fusion. Science. 1980;**209**: 713-715

[226] Hughes RN. A Functional Biology of Clonal Animals. New York: Chapman and Hall; 1989

[227] Hall VR, Hughes TP. Reproductive strategies of modular organisms: Comparative studies of reef-building corals. Ecology. 1996;**77**:950-963

[228] Wood R. Reef Evolution. Oxford: Oxford University Press; 1999

[229] de Weerdt WH. Transplantation experiments with Caribbean *Millepora* species (Hydrozoa, Coelenterata), including some ecological observations on growth forms. Bijdragen tot de Dierkunde. 1981;**51**:1-19

[230] Witman JD. Effects of predation by the fireworm *Hermodice carunculata* on milleporid hydrocorals. Bulletin of Marine Science. 1988;**42**(3):446-458

[231] Rahav O, Ben-Zion M, Achituv Y, Dubinsky Z. A photographic, computerized method for in situ growth measurements in reef-building cnidarians. Coral Reefs. 1991;**9**(4):204

[232] Lewis JB. Banding, age and growth in the calcareous hydrozoan *Millepora complanata* Lamarck. Coral Reefs. 1991; **9**(4):209-214

[233] Glynn PW. *Acanthaster*: Effect on coral reef growth in Panama. Science. 1973;**180**(4085):504-506

[234] Schweinsberg M, Tollrian R, Lampert KP. Inter- and intra-colonial genotypic diversity in hermatypic hydrozoans of the family Milleporidae. Marine Ecology. 2017;**38**(1):e12388

[235] Dubé CE, Planes S, Zhou Y, Berteaux-Lecellier V, Boissin E. On the occurrence of intracolonial genotypic variability in highly clonal populations of the hydrocoral *Millepora platyphylla* at Moorea (French Polynesia). Scientific Reports. 2017;**7**(1):14861

[236] Puill-Stephan E, van Oppen MJH, Pichavant-Rafini K, Willis BL. High potential for formation and persistence of chimeras following aggregated larval settlement in the broadcast spawning coral, *Acropora millepora*. Proceedings of the Royal Society of London B: Biological Sciences. 2012;**279**:699-708

[237] Amar KO, Chadwick NE, Rinkevich B. Coral kin aggregations exhibit mixed allogeneic reactions and enhanced fitness during early ontogeny. BMC Evolutionary Biology. 2008;**8**:126

[238] Puill-Stephan E, Willis BL, Abrego D, Raina J-B, van Oppen MJH. Allorecognition maturation in the broadcast spawning coral *Acropora millepora*. Coral Reefs. 2012;**31**: 1019-1028

[239] D'Aloia CC, Bogdanowicz SM, Majoris JE, Harrison RG, Buston PM. Self-recruitment in a Caribbean reef fish: A method for approximating dispersal kernels accounting for seascape. Molecular Ecology. 2013;**22**: 2563-2572

[240] Cuif M, Kaplan DM, Fauvelot C, Lett C, Vigliola L. Monthly variability of

self-recruitment for a coral reef damselfish. Coral Reefs. 2015;**34**: 759-770

[241] Galzin R, Pointier JP. Moorea Island, society archipelago. In: Proceedings of the Fifth International Coral Reef Symposium. Vol. 1. 1985. pp. 73-102

[242] Kayal M, Vercelloni J, De Loma TL, Bosserelle P, Chancerelle Y, Geoffroy S, et al. Predator crown-of-thorns starfish (*Acanthaster planci*) outbreak, mass mortality of corals, and cascading effects on reef fish and benthic communities. PLoS One. 2012;7(10): e47363

[243] Holbrook SJ, Adam TC, Edmunds PJ, Schmitt RJ, Carpenter RC, Brooks AJ, et al. Recruitment drives spatial variation in recovery rates of resilient coral reefs. Scientific Reports. 2018;**8**(1):7338

[244] Bengtsson BO. Genetic variation in organisms with sexual and asexual reproduction. Journal of Evolutionary Biology. 2003;**16**:189-199

[245] Harrison PL, Wallace CC. Reproduction, dispersal and recruitment of scleractinian corals. Ecosystems of the World. 1990;**25**: 133-207

[246] Harrison PL, Booth DJ. Coral reefs: Naturally dynamic and increasingly disturbed ecosystems. In: Marine Ecology, Connell SD and Gillanders BM editors. Melbourne: Oxford University Press; 2007. pp. 316-377

[247] Courchamp F, Clutton-Brock T, Grenfell B. Inverse density dependence and the Allee effect. Trends in Ecology & Evolution. 1999;**14**:405-410

[248] Randall RH, Cheng YM. Shallow water hydrozoan corals. Acta Geologica Taiwanica. 1984;**22**:35-99

[249] Bellwood DR, Hughes TP, Folke C, Nyström M. Confronting the coral reef crisis. Nature. 2004;**429**:827-833

[250] Bruno JF, Selig ER. Regional decline of coral cover in the indo-Pacific: Timing, extent, and subregional comparisons. PLoS One. 2007;**2**:e711

[251] De'ath G, Fabricius KE, Sweatman H, Puotinen M. The 27–year decline of coral cover on the great barrier reef and its causes. Proceedings of the National Academy of Sciences of the United States of America. 2012;**109**: 17995-17999

[252] Orr JC, Fabry VJ, Aumont O, Bopp L, Doney SC, Feely RA, et al. Anthropogenic Ocean acidification over the twenty-first century and its impact on calcifying organisms. Nature. 2005; **437**(7059):681

[253] Norström AV, Nyström M, Lokrantz J, Folke C. Alternative states on coral reefs: Beyond coral–macroalgal phase shifts. Marine Ecology Progress Series. 2009;**376**:295-306

[254] Glynn PW, Maté JL, Baker AC, Calderón MO. Coral bleaching and mortality in Panama and Ecuador during the 1997–1998 El Niño-southern oscillation event: Spatial/temporal patterns and comparisons with the 1982–1983 event. Bulletin of Marine Science. 2001;**69**:79-109

[255] Brown D, Edmunds PJ. Long-term changes in the population dynamics of the Caribbean hydrocoral *Millepora* spp. Journal of Experimental Marine Biology and Ecology. 2013;**441**:62-70

Chapter 3

Chemoprotective Effect of Edible Gastropod, *Xancus pyrum* and Its Usefulness in the Amelioration of Cisplatin Induced Toxicity

Jayaprakash Bindhu and Das Arunava

Abstract

The main purpose of this study was to evaluate chemoprotective activities of methanolic extracts of an edible gastropod (*Xancus pyrum*) in cisplatin-induced immunosuppressed mice. Cisplatin (100 mg/kg, intraperitoneally [IP]) induced immunosuppressed mice were treated with a methanolic extract of *X. pyrum* (0.5 mg/dose/animal/IP) for a period of 10 days. The effect of the extract on lymphoid organ weight, bone marrow cellularity (BMC), alpha esterase activity, and on enzyme levels such as serum glutamic oxaloacetic transaminase, serum glutamic pyruvic transaminase, urea, and creatinine was estimated to identify the chemoprotective activity of *X. pyrum*. The administration of *X. pyrum* extract in cisplatin-treated mice, found to enhance the BMC and alpha-esterase positive cells, which were drastically reduced in cisplatin alone treated control animals suggests that cisplatin-induced myelosuppression was reversed or inhibited by *X. pyrum* extract administration possibly through its chemoprotective activity. In conclusion, cisplatin and its metabolites can bind to DNA, causing damage that may result in chromosome breaks, micronucleus formation and cell death. Administration of *X. pyrum* extract in cisplatin-treated mice, found to enhance the BMC and alpha-esterase positive cells, which were drastically reduced in cisplatin alone treated control animals suggests that cisplatin-induced myelosuppression was reversed or inhibited by *X. pyrum* extract administration possibly through its chemoprotective activity.

Keywords: gastropod, *Xancus pyrum*, chemoprotective, cisplatin, alpha esterase

1. Introduction

Cancer is one of the leading causes of death around the world; it is characterized with the aid of uncontrolled growth and unfold of odd cells. If the unfold isn't managed, it can bring about demise. Cancer is resulting from both external components (tobacco, chemical compounds, radiation and infectious organisms) and internal factors (inherited mutations, hormones, immune conditions and mutations that arise from metabolism). Those casual factors may additionally act together in series to initiate or promote carcinogenesis [1]. Malignancy is a perplexing arrangement of illnesses. Every malignancy is exceptional in the manner it develops and builds up, its odds of spreading, the manner in which it influences one's body and the side effects one may involvement. A few components, including area and how the destructive cells show up under the magnifying instrument, decide how disease is analysed. All malignant growths, be that as it may, can be categorized as one of four general classes, for example, carcinoma, leukaemia, sarcoma and lymphoma. Carcinoma is a type of cancer developed from epithelial cells. This is the single largest group of human cancers forming about 80% of all cancers. Tumours arising from connective tissue cells (mesenchymal tissue) such as fibroblasts or bone cells are called sarcomas. Cancers of blood forming cells are called leukaemia. Sometimes cancer produced from the lymphoid origin is localized in lymph glands called lymphomas [2]. Cancer is treated with surgery, radiation, chemotherapy, hormone therapy, biological therapy and targeted therapy. Radiotherapy and chemotherapy remain the dominant weapons in the arsenal for the treatment of cancer. They kill not only the tumour cells but also normal cells [3]. An enormous number of tumours are ineffectively responsive or even non-reacting to remedial medications and radiotherapy. Expanding the dosages of cytotoxic medications and radiation neglect to improve the reaction to these treatments and a significant number of them show protection from slaughtering. A perfect methodology is distinguish anticancer specialists that trigger adequately the procedure of cell demise specially in tumour cells [4]. Radiation and chemotherapy treatments are made as they wipe out fast growing cancer cells, they can also damage fast growing normal cells. They can affect some healthy, fast growing cells causing side effects some of the most common include, Depression of the immune system, which can result in potentially fatal infections. Fatigue, the treatment can be physically exhausting for the patient, who might already be very tired from cancer-related fatigue. It may produce mild to severe anaemia. Inclination to drain effectively, prescriptions that murder quickly separating cells or platelets are probably going to lessen the quantity of platelets in the blood, which can result in wounds and dying. Amazingly low platelet checks might be incidentally helped through platelet transfusions. At times, chemotherapy medicines are delayed to permit platelet checks to recuperate. Gastrointestinal misery, nausea and regurgitating are basic symptoms of chemotherapeutic drugs that murder quick partitioning cells. This can likewise deliver loose bowels or blockage, malnutrition and lack of hydration. This can result in fast weight reduction. Male pattern baldness, a few prescriptions that slaughter quickly isolating cells cause sensational male pattern baldness; different drugs may make hair meager. These are transitory impacts: hair for the most part begins developing back half a month after the last treatment, now and again with an inclination to twist that might be known as a "chemo perm" [5].

Natural products have a long history of use in the service of mankind for the prophylaxis and treatment of several diseases and cancer is not an exception. For such a dreadful disease, apart from conventional modalities like surgery, radiotherapy and chemotherapy a few other approaches are also available or being tried. It has been reported that one-third of cancer patients use some form of complementary and other alternative medicines. In the recent times considerable attention has been focused on the identification and development of natural products for chemoprevention by systemic and rigorous screening processes many of the potential chemo preventive agents have shown considerable safety and efficacy in preclinical evaluation and are in the stages of clinical testing [6]. The class **Gastropoda** or **gastropods** (also previously known as univalves sometimes also spelled Gasteropoda) form a major part of the phylum Mollusca. Gastropods are more commonly known as snail and slugs. This is the most diversified class in the phylum with 60,000–80,000 living species. There are 409 recent families of gastropods. Fossil gastropods represent another 202 families. This class of animals is second only to insects in its number of known species. The gastropods include many thousands of species of marine snails and sea slugs, as well as freshwater snails and fresh water limpets, and the terrestrial (land) snails and slugs [7].

The marine environment is a rich wellspring of both organic and compound biodiversity which has been investigated in the revelation of one of a kind synthetic compounds, having potential for mechanical improvement as pharmaceuticals, makeup, wholesome enhancements, atomic tests, fine synthetics and agrochemicals. As of late of novel metabolites with powerful pharmacological properties have been found from marine living being. One among them is gastropod which incorporates gastropod *Xancus pyrum* [8]. *Xancus pyrum* Linnaeus (Xancidae, Gastropoda) vernacular name Sankh shell, the moderate moving creature. The marine oils present uncommon troubles in the examination in view of the wide assortment of unsaturated fats. Though normal oils may for the most part be dissected as far as individual acids, if there should arise an occurrence of marine oils it is just conceivable to assess the different acids as per chain length. The fatty acids from the gastropod *Xancus pyrum* were obtained through extraction, isolation and chromatographic separation of visceral mass of the animals. Of the 14 fatty acids methyl esters investigated 8 were saturated fatty acids and 6 unsaturated fatty acids. Whereas out of 8 saturated fatty acids, 5 of them were the common acids. These fatty acids were used to treat cardiac diseases and obesity [9].

2. Taxonomic position of *Xancus pyrum*

2.1 Gastropod description

The word gastropod is from the Greek, gastro importance stomach and poda significance foot, thus stomach-foot, a somewhat human name dependent on the way that to people it appears that snails and slugs slither on their tummies. In actuality, snails and slugs have all their viscera, including their stomach, in a mound on the inverse, dorsal side of the body. The shell has extensive noteworthiness in Hinduism and Buddhism. It is viewed as hallowed and is one of the eight propitious images. In these religious settings the shell is now and again changed by having the tip of the tower cut off, with the goal that it tends to be blown as a formal trumpet. A few shells that are utilized along these lines are brightened with metal and semiprecious stones. The gastropod shell is the piece of the body of a gastropod or snail, a sort of mollusk. The gastropod shell is an outside skeleton or exoskeleton, which serves for muscle connection, yet additionally for security from predators and from mechanical harm. In land snails the shell is a basic insurance against the sun, and against drying out. The gastropod shell has a few layers, and is commonly made of calcium carbonate accelerated out into a natural framework known as conchiolin. The shell is discharged by a piece of the molluscan body known as the mantle. Not all gastropods have a shell, however the greater part do. In pretty much every case the shell comprises of one piece, and is ordinarily spirally snaked, albeit a few gatherings, for example, the different various families and genera of limpets, have basic cone-formed shells as grown-ups [10].

2.2 *Xancus pyrum*

Xancus pyrum, common names the chank shell, sacred chank or chank, also known as the divine conch, sometimes referred to simply as a conch, is a species of very large sea snail with a gill and an operculum. It is a marine gastropod mollusk in the family Turbinellidae. This species occurs in the Indian Ocean. The name "chank" for the shell of this species is derived from the Indian word shankha, the divine conch. The old generic name was Turbinella. The Dutch used to call them chianco.

The shell of this species is massive, with three or four prominent columellar plicae. It is usually pure white under a heavy brown periostracum, but it can also be a pale apricot colour. It can sometimes be dotted with dark brown. The shell has impressive importance in Hinduism and Buddhism. It is viewed as hallowed and is of the eight propitious images. In these religious settings the shell is now and again altered by having the tip of the tower cut off, with the goal that it tends to be blown as a formal trumpet. A few shells that are utilized along these lines are beautified with metal and semiprecious stones. The shell of this species is quite often right - turned in its shell-looping, yet all around once in a while a left-gave, or sinistral, shell is found. In the Hindu religious setting, the exceptionally uncommon left-gave or sinistral shells of this species are known as "Dakshinavarti", instead of the more typical right-gave ones which are known as "Vamavarti". The Dakshinavarti is especially exceptionally esteemed as far as its religious importance.

Systematic position:

Kingdom: Animalia
Phylum: Mollusca
Class: Gastropoda
Clade: Neogastropoda
Superfamily: Muricoidea
Family: Turbinellidae
Subfamily: Turbinellinae
Genus: *Xancus*
Species: *X. pyrum*

3. Methodology

3.1 Animals

Innate BALB/C (6 two months) mice, gauging 23–28 g, were gotten from Pasteur Institute, rearing area, Coonoor. The creatures were housed in ventilated plastic enclosures at 37$\pm$ 1°c, 40 $\pm$ 10% dampness, and 12/12-h light/dull cycles during about 14 days of acclimatization to research facility conditions and all through the whole test time frame. The creatures were encouraged with ordinary mouse chow (Sai Feeds, Mumbai, India), and given water ad libitum. Every single creature investigation was directed by the guidelines and guidelines of Animal Ethics Committee, Government of India.

3.2 Preparation and administration of extract

3.2.1 Preparation of extract

Xancus pyrum (Gastropod) were collected from Shell meat dealers, Tuticorin, South east coast of India. The Gastropod meat was washed in distilled water and dried in a hot air oven at 50°C. The dried meat was powdered and extracted overnight by stirring with 10 volumes of 75% methanol. Supernatant was collected after centrifuging at 3000 rpm for 10 min. The solvent was evaporated to dryness at 45°C in hot water bath. The yield of the extract was 10%.

3.2.2 Administration of extract

For animal administration the extract was dissolved in minimum quantity of methanol, then resuspended in 1% gum acacia in phosphate buffered saline and given at a concentration of 0.5 g/dose/animal/intraperitoneally. For in vitro experiments, the extract was dissolved in dimethyl sulfoxide (DMSO) and diluted in the medium so that the concentration of DMSO was less than 0.1%vol/vol.

3.3 Chemoprotective effect of edible gastropod meat, *Xancus pyrum*

3.3.1 Experimental protocol

The animals were divided into three groups of six animals each as follows:

Group 1: Normal animals, without any treatment.

Group 2: Treated animals received cisplatin alone dissolved in 1% gum acacia intraperitoneally for 10 consecutive days.

Group 3: Treated animals received *Xancus pyrum* (0.5 mg) methanolic extracts. Dissolved in 1% gum acacia intraperitoneally for 10 consecutive days.

3.3.2 Determination of the effects of on Xancus pyrum on lymphoid organ weight in cisplatin treated animals

Eighteen animals were randomly divided into 3 groups containing six animals each, one as normal, which did not receive any treatment. The second group as treated animals, treated with cisplatin alone. Third group treated with cisplatin and *Xancus pyrum* Three animals from each group were sacrificed at two different time intervals (7th and 11th day) by cervical dislocation. Body weight of each animal was taken before sacrifice, lymphoid organs such as thymus and spleen was excised, weighed and expressed as relative organ weight.

3.3.3 Determination of the effects of Xancus pyrum on bone marrow cellularity (BMC) in cisplatin treated animals

Bone marrow cellularity was done according to the method. Bone marrow was collected from the femur into the medium containing 2% serum and made into single cell suspension. The number of cells was determined using a haemocytometer and expressed as total cells determined by tryphan blue (1% in saline) exclusion method per femur [11].

3.3.4 Determination of the effects of Xancus pyrum on alpha esterase activity in cisplatin treated animals (azodye coupling method, Bancroft and Cook, 1984)

3.3.4.1 Principle

Esterase enzyme present or absent in monocytes hydrolyses the substrate alpha-naphthyl acetate to form an invisible primary reaction product (PRP). The complex is coupled with the diazonium salt to produce coloured final reaction product under the microscope [12].

3.3.4.2 Reagents

1. **Fixative solution.** Formaldehyde (HCHO 37%, 25 ml) acetone (45 ml), double distilled water (30 ml), disodium hydrogen phosphate (Na_2HPO_4-20 mg) and potassium dihydrogen phosphate (KH_2PO_4-100 mg).

Solution A. Pararosaline was prepared by dissolving 1 g powder in 20 ml double distilled water and 5 ml HCl. Gently warmed solution is filtered and stored in dark at 4°C.

2. **Solution B.** Sodium nitrate (4%) was prepared by dissolving 400 mg in 10 ml double distilled water.

3. **Solution C.** Alpha naphthyl acetate was prepared by 50 mg powder in 2.5 ml glycol monoethyl ether.

4. **Phosphate buffer** (pH 7.4)

5. **Harris haematoxylin.** (500 ml) was prepared by dissolving 2.5 g powder in 50 ml ethyl glycol added to a day before prepared supersaturated solution of alu (90 g in 500 ml double distilled water) and sodium iodide or potassium iodide (20 mg). The stain was stirred overnight, filtered and stored in a dark bottle.

3.3.4.3 Procedure

Bone marrow from both femurs of mice was collected in PBS, washed thrice and smeared over the slides. Air dried slides were fixed in freshly prepared fixative 30 s at 4°C and dipped in double distilled water thrice. Air dried slides were incubated at room temperature in the following freshly prepared filtered solution. 1.2 ml solution A and 1.2 ml solution B was mixed well and allowed to react for 1 min after which solution C was added and was made up to 50 ml solution by phosphate buffer (pH 7.4).

Slides were incubated in above solution for 45 min at 37°C. After incubation slides were washed in double distilled water for 10 min and counter stained with haematoxylin for 1 min. After staining slides were washed in water for long time and observed under microscope (100×, oil immersion) for scoring positive and negative alpha esterase cells out of 4000 cells.

3.4 Determination of the effect of *Xancus pyrum* on enzyme levels in cisplatin treated animals

Liver homogenates were made in ice cold Tris buffer (0.1 M pH 7.4) and was used for the estimation of SGOT, SGPT, urea and creatinine. Serum was also used to estimate all the above parameters.

3.4.1 Estimation of SGPT (Span Diagnostics Ltd., Surat, India)

3.4.1.1 Principle

Alanine aminotransferase (ALT) catalyses the transamination of L-alanine and α-ketoglutarate to form pyruvate and L-glutamate. Pyruvate so formed is coupled with 2,4-dinitrophenyl hydrazine (2,4-DNPH) to form a corresponding hydrazone, a brown coloured complex in alkaline medium and this can be measured colorimetrically [13].

$$\text{L} - \text{alanine} + \alpha - \text{ketoglutarate} \rightleftarrows \text{pyruvate} + \text{L} - \text{glutamate} \quad (1)$$

$$\text{Pyruvate} + 2,4 - \text{DNPH} \rightleftarrows \text{corresponding hydrazone (brown colour)} \quad (2)$$

Reitman and Frankel method is an end-point colorimetric method for the estimation of enzyme activity.

3.4.1.2 Reagents

Reagent no.	Reagents	Composition
1.	Buffered alanine-α-KG substrate, pH 7.4	L-alanine α-KG Phosphate buffer Preservative Stabilizer
2.	2,4-DNPH Colour Reagent	2,4-Dinitrophenyl hydrazine Preservative Stabilizer
3.	Sodium hydroxide, 4 N	Sodium hydroxide
4.	Working pyruvate standard, 8 mM (150 IU/L)	Sodium pyruvate Preservative Stabilizer

Working reagent preparation:

- Reagent 1, 2 and 4 are ready to use.
- Solution I: Dilute 1 mL of Reagent 3 to 10 mL with purified water.

3.4.1.3 Procedure

Pipette into tube marked	Blank	Standard	Test	Control
Volume in mL				
Reagent 1	0.25	0.25	0.25	0.25
Serum	—	—	0.05	—
Standard	—	0.05	—	—
Mix well and incubate at 37°C for 30 min				
Reagent 2	0.25	0.25	0.25	0.25
Deionized water	0.05	—	—	—
Serum	—	—	—	0.05
Mix well and allow to stand at room temperature (15–30°C) for 20 min				
Solution I	2.5	2.5	2.5	2.5

Mix well and read the optical density against purified water in a photometer at 505 nm, within 15 min.

Calculation:

$$\text{AST (GPT) activity (IU/L)} = \frac{\text{absorbance of test} - \text{absorbance of control} \times \text{concentration of standard}}{\text{absorbance of standard} - \text{absorbance of blank}} \tag{3}$$

where concentration of standard = 150 IU/L.

3.4.2 Estimation of SGOT (Span Diagionostics Ltd., Surat, India)

3.4.2.1 Principle

Aspartate aminotranferase (AST) catalyses the transamination of L-aspartate and α-ketoglutarate to form L-glutamate and oxaloacetate. Oxaloacetate so formed is coupled with 2,4-dinitrophenyl hydrazine (2,4-DNPH) to form a corresponding hydrazone, a brown coloured complex in alkaline medium and this can be measured colorimetrically [14].

$$\text{L} - \text{aspartate} + \alpha - \text{ketoglutarate} \rightleftarrows \text{oxaloacetate} + \text{L} - \text{glutamate} \quad (4)$$

$$\text{Oxaloacetate} + 2,4 - \text{DNPH} \rightleftarrows \text{corresponding hydrazone (brown colour)} \quad (5)$$

3.4.2.2 Reagents

Reagent no.	Reagents	Composition
1.	Buffered aspartate-α-KG substrate, pH 7.4	L-aspartic acid α-KG Phosphate buffer Preservative Stabilizer
2.	2,4-DNPH colour reagent	2,4-dinitrophenyl hydrazine Preservative Stabilizer
3.	Sodium hydroxide, 4 N	Sodium hydroxide
4.	Working pyruvate standard, 6 mM (114 IU/L)	Sodium pyruvate Preservative Stabilizer

Working reagent preparation:

- Reagent 1, 2 and 4 are ready to use.
- Solution I: Dilute 1 mL of Reagent 3 to 10 mL with purified water.

3.4.2.3 Procedure

Pipette into tube marked	Blank	Standard	Test	Control
Volume in mL				
Reagent 1	0.25	0.25	0.25	0.25
Serum	—	—	0.05	—
Standard	—	0.05	—	—
Mix well and incubate at 37°C for 30 min				
Reagent 2	0.25	0.25	0.25	0.25
Deionized water	0.05	—	—	—
Serum	—	—	—	0.05
Mix well and allow to stand at room temperature (15–30°C) for 20 min				
Solution I	2.5	2.5	2.5	2.5

Mix well and read the optical density against purified water in a photometer at 505 nm, within 15 min.

Calculation:

$$\text{AST (GOT) activity (IU/L)} = \frac{\text{absorbance of test} - \text{absorbance of control} \times \text{concentration of standard}}{\text{absorbance of standard} - \text{absorbance of blank}} \quad (6)$$

where concentration of standard = 114 IU/L.

3.5 Effect of *Xancus pyrum* on the biochemical parameters after cisplatin administration

3.5.1 Estimation of Urea (Span Diagnostics Ltd., Surat, India)

3.5.1.1 Principle

Urea is converted quantitatively to ammonia and CO_2 in the presence of urease. The ammonium ions react with the phenolic chromogen and hypocrite to give a green coloured complex. The intensity of the colour formed is measured at 578 nm and is directly proportional to the concentration of urea in test specimen [15].

$$\text{Urea} + H_2O \rightarrow \text{ammonia} + CO_2 \quad (7)$$

$$\text{Ammonia} + \text{phenolic chromogen} + \text{hypochlorite} \rightarrow \text{green coloured complex} \quad (8)$$

3.5.1.2 Reagents

Reagent 1: Urea (Enzyme reagent).
Reagent 2: Urea (Chromogen reagent).
Reagent 3: Urea diluents (buffer).
Reagent 4: Urea standard (50 mg/dL).

3.5.1.3 Working Reagent preparation

i. Transfer the contents of one vial of Reagent 1 to the bottle of Reagent 3.

ii. Rinse 1 Urea vial properly with 3 Urea diluents and mix gently.

iii. Store these working reagents at 2–8°C when not in use.

iv. 2 Urea reagent and Urea standard are ready to use.

3.5.1.4 Procedure

Pipette into clean and dry test tubes labelled as blank (B), standard (S) & test (T):

Addition sequence	Blank	Standard	Test
Working reagent	1000 μL	1000 μL	1000 μL
Distilled water	10 μL	—	—

Addition sequence	Blank	Standard	Test
Urea standard	—	10 μL	—
Specimen	—	—	10 μL
Mix well and incubate for 5 min at 37°C or 15 min at room temperature			
2-Urea reagent	1000 μL	1000 μL	1000 μL

1. Mix well and incubate for 5 min at 37°C or 15 min at room temperature.
2. Measure the absorbance of standard (Abs. S) & test sample (Abs. T) against the blank at 578 nm.
3. Colour is stable for 45 min when protected from light, so absorbance should be measured within that period.

3.5.1.5 Calculation

$$\text{Urea concentration in mg/dL of test specimen} = \text{Abs T} + \text{Abs S} \times 50 \tag{9}$$

$$\text{Blood urea nitrogen (BUN) in mg/dL} = \text{urea (in mg/dL)} \times 0.467 \tag{10}$$

3.5.2 Estimation of creatinine (Span Diagnostic Ltd., Surat, India)

3.5.2.1 Principle

Creatinine responds with picric corrosive in antacid medium to frame an orange hued complex. The rate of arrangement of this complex is estimated by perusing the adjustment in absorbance at 520 nm in a chose interim of time and is relative to the centralization of creatinine. The response time and the grouping of picric corrosive and sodium hydroxide have been streamlined to maintain a strategic distance from obstruction from keto acids [16].

$$\text{creatinine} + \text{picric acid} \rightarrow \text{orange coloured complex} \tag{11}$$

3.5.2.2 Reagents

Reagent 1: Picric acid.
Reagent 2: Sodium hydroxide 0.75 N
Reagent 3: Stock Creatinine Standard, 150 mg

3.5.2.3 Working solution preparation

Dilute 0.1 mL of Reagent 3 to 10 mL with purified water and mix well.

3.5.2.4 Procedure

Step A. Deproteinization of test sample

Serum/plasma	0.5 mL
Purified water	0.5 mL
Reagent 1: Picric acid	3.0 mL

Mix well, keep in a boiling water bath exactly for 1 min. Cool immediately under running tap water and centrifuge or filter.
Step B. Colour development.

	Blank (B)	Standard (S)	Test (T)
Filtrate/supernatant (from Step A)	—	—	2.0 mL
Working standard	—	0.5 mL	—
Purified water	0.5 mL	—	—
Reagent 1: Picric acid	1.5 mL	1.5 mL	—
Reagent 2: sodium hydroxide, 0.75 N	0.5 mL	0.5 mL	0.5 mL

Mix well, allow to stand at room temperature exactly for 20 min and measure immediately the optical density of blank (B), standard (S) and test (T) against purified water at 520 nm.

Calculation:

$$\text{serum creatinine in mg/100 mL} = \frac{\text{O.D.test} - \text{O.D.blank} \times 3}{\text{O.D.std.–O.D.blank}} \tag{12}$$

3.6 Statistical analysis

The results are expressed in mean ± standard deviation (SD). Statistical analysis was performed by using Students 't 'test. p values < 0.05 were considered to be statistically significant.

4. Results and discussion

4.1 Chemoprotective effect of edible gastropod meat

4.1.1 Xancus pyrum

4.1.1.1 Effect of Xancus pyrum on relative organ weights after cisplatin administration

Body weight of each animal was taken before sacrifice, lymphoid organs such as thymus and spleen was excised, weighed and expressed as relative organ weight.

The cisplatin treated animals showed high reduction in the weight of all the organs such as 0.23 ± 0.02 g/100 g body weight of spleen, 0.17 ± 0.01 g/100 g body weight of thymus, 3.7 ± 0.19 g/100 g body weight of liver, 1.2 ± 0.01 g/100 g body weight of kidney, 0.62 ± 0.01 g/100 g body weight of lungs. The cisplatin treated along with *Xancus pyrum* mice showed a significantly increase in the weight of all the organs such as 0.34 ± 0.072 g/100 g body weight of spleen, 0.23 ± 0.01 g/100 g body weight of thymus, 4.84 ± 0.05 g/100 g body weight of liver, 1.37 ± 0.18 g/100 g body weight of kidney, 0.769 ± 0.05 g/100 g body weight of lungs.

The p values of cisplatin treated animals along with *Xancus pyrum* for spleen was $p < 0.01$, which was less significant but for thymus, liver, kidney and lungs it was $p < 0.05$ which was considered to be statistically significant.

Weight of all relative organs was increased in cisplatin treated animals by the extract administration, providing supportive evidence for *Xancus pyrum* extract is immunostimulative. The results are given in **Table 1**.

Treatment	Relative organ weight (g/100 g body weight)									
	Spleen		Thymus		Liver		Kidney		Lungs	
	7th day	11th day	7th day	11th day	7th day	11th day	7th day	11th day	7th day	11th day
Normal	0.50 ± 0.11	0.69 ± 0.02	0.16 ± 0.04	0.18 ± 0.02	5.41 ± 0.38	5.98 ± 0.39	1.27 ± 0.17	1.26 ± 0.22	0.54 ± 0.03	0.59 ± 0.02
Cisplatin alone	0.23 ± 0.02	0.20 ± 0.04	0.17 ± 0.01	0.15 ± 0.07	3.78 ± 0.19	3.60 ± 0.40	1.23 ± 0.01	0.98 ± 0.04	0.62 ± 0.01	0.53 ± 0.03
Cisplatin + *Xancus pyrum*	0.34 ± 0.07**	0.41 ± 0.14**	0.23 ± 0.01**	0.26 ± 0.04*	4.84 ± 0.05*	5.58 ± 0.30*	1.37 ± 0.18*	1.43±0.08**	0.76 ± 0.05*	0.79 ± 0.08*

Values are expressed as mean ± SD.
*$p < 0.05$.
**$p < 0.01$.
***$p < 0.001$.

Table 1.
Effect of Xancus pyrum on relative organ weights in cisplatin treated animals.

4.1.1.2 Effect of Xancus pyrum on bone marrow cellularity and α-esterase activity after cisplatin administration

Bone marrow was collected from the femur into the medium containing 2% serum and made into single cell suspension. The number of cells was determined using a haemocytometer and expressed as total cells determined by tryphan blue (1% in saline) exclusion method per femur.

Effect of *Xancus pyrum* on bone marrow cellularity and α-esterase activity is given in **Table 2**. The number of bone marrow cells as well as α-esterase positive cells was decreased drastically in cisplatin alone treated animals, but this was significantly ($p < 0.001$) reversed by administration of *Xancus pyrum*. In cisplatin treated animals, on the 7th day there was a drastic reduction in the number of bone marrow cells ($25.5 \times 10^5 \pm 1.414$ cells/femur) and α-esterase positive cells (634.5 ± 3.05 positive cells/4000 cells) compared to *Xancus pyrum* treated along with cisplatin animals. Treatment with *Xancus pyrum* could elevate the bone marrow cellularity and number of α-esterase positive cells. In cisplatin treated group of animals along with *Xancus pyrum*, bone marrow cellularity and α-esterase positive cells was found to be $68.30 \times 10^5 \pm 4.24$ cells/femur and 1179 ± 2.121 cells/4000 bone marrow cells respectively on 7th day and it was again enhanced to $69.7 \times 10^5 \pm 4.24$ cells/femur and 1227 ± 1.414 cells/4000 bone marrow cells on 11th day respectively compare to the cisplatin alone treated animals ($20.93 \times 10^5 \pm 3.055$ cells/femur and 620.66 ± 3.05 cells/4000 bone marrow cells).

4.1.1.3 Effect of Xancus pyrum on enzyme levels after cisplatin administration

4.1.1.3.1 Serum glutamic oxaloacetic transaminase (SGOT)

An enzyme that is normally present in liver and heart cells. SGPT is released into blood when the liver or heart are damaged. The blood SGPT levels are thus elevated. Also called aspartate aminotransferase (AST).

A significant increase in the levels of SGOT (82.280 ± 2.7 IU/L) and SGPT (85.22 ± 2.393 IU/L) observed in the serum samples of cisplatin alone treated group was reversed by the administration of *Xancus pyrum*. Treatment with cisplatin along with *Xancus pyrum* significantly reduced the levels of SGOT (52.68 ± 0.46 IU/L) and SGPT (55.820 ± 1.814 IU/L) in serum, that is the p value was found to be $p < 0.001$ showing that the extract is highly significant.

Treatment	Bone marrow cellularity (cells/femur)		α-Esterase activity (no. of α-esterase positive cells/4000 cells)	
Days	**7th day**	**11th day**	**7th day**	**11th day**
Normal	$85.0 \times 10^5 \pm 2.828$	$89.5 \times 105 \pm 3.536$	884 ± 2.828	892.5 ± 2.121
Cisplatin alone	$25.5 \times 10^5 \pm 1.414$	$20.9 \times 10^5 \pm 3.055$	634.5 ± 3.055	620.66 ± 3.055
Cisplatin + *Xancus pyrum*	$68.30 \times 10^5 \pm 4.242^{***}$	$69.7 \times 10^5 \pm 4.950^{***}$	$1179 \pm 2.121^{***}$	$1227 \pm 1.414^{***}$

Values are expressed as mean ± SD.
$^{*}p < 0.05$.
$^{**}p < 0.01$.
$^{***}p < 0.001$.

Table 2.
Effect of Xancus pyrum on bone marrow cellularity and α-esterase activity in cisplatin treated animals.

4.1.1.3.2 Serum glutamic pyruvic transaminase (SGPT)

An enzyme that is normally present in liver and heart cells. SGPT is released into blood when the liver or heart is damaged. The blood SGPT levels are thus elevated. Also called alanine aminotransferase (ALT).

Cisplatin treated animals showed a decrease in the levels of SGOT (32.67 ± 2.7 IU/L) and SGPT (42.04 ± 1.9 IU/L) observed in the liver sample. Administration of *Xancus pyrum* significantly increased the level of SGOT (41.545 ± 1.3 IU/L) and SGPT (48.290 ± 1.4 IU/L) in liver. The levels of SGOT and SGPT values are given in **Tables 3** and **4**. The SGPT level was increased drastically in cisplatin alone treated animals, but this was significantly ($p < 0.001$) reversed by administration of *Xancus pyrum* extract.

4.1.1.4 Effect of Xancus pyrum on the biochemical parameters after cisplatin administration

The blood urea nitrogen (BUN) test is a measure of the amount of nitrogen in the blood in the form of urea, and a measurement of renal function. **Urea** is a substance secreted by the liver, and removed from the blood by the kidneys.

The renal functions can be estimated by biochemical parameters like BUN (blood urea nitrogen) and creatinine is given in **Tables 5** and **6**. Cisplatin administration in mice was found to increase the BUN concentration in serum on 7th day 18.19 ± 0.2 mg/dL and 11th day 2.20 ± 0.04 mg/dL but this was significantly reduced to 7.00 ± 0.12 mg/dL on 7th day and 7.025 ± 0.05 mg/dL on 11th day by

Group	Serum GOT (IU/L)		Liver GOT (IU/L)	
Days	7th day	11th day	7th day	11th day
Normal	9.064 ± 0.3	9.172 ± 0.1	90.16 ± 2.9	97.41 ± 2.1
Cisplatin alone	76.92 ± 1.8	82.28 ± 2.7	43.88 ± 2.3	32.67 ± 2.7
Cisplatin + *Xancus pyrum*	55.93 ± 0.1***	52.68 ± 0.4***	39.85 ± 0.7**	41.54 ± 1.3**

Values are expressed as mean ± SD.
$^{*}p < 0.05$.
$^{**}p < 0.01$.
$^{***}p < 0.001$.

Table 3.
Effect of Xancus pyrum treatment on the serum, liver SGOT levels in cisplatin treated animals.

Group	Serum GPT (IU/L)		Liver GPT (IU/L)	
Days	7th day	11th day	7th day	11th day
Normal	9.390 ± 0.2	9.290 ± 0.04	69.45 ± 2.05	70.36 ± 0.09
Cisplatin alone	74.65 ± 1.4	85.22 ± 2.3	40.91 ± 0.96	32.04 ± 1.9
Cisplatin + *Xancus pyrum*	58.22 ± 0.3**	55.82 ± 1.8***	45.22 ± 0.9**	48.29 ± 1.4***

Values are expressed as mean ± SD.
$^{*}p < 0.05$.
$^{**}p < 0.01$.
$^{***}p < 0.001$.

Table 4.
Effect of Xancus pyrum treatment on the serum, liver SGPT levels in cisplatin treated animals.

the administration of *Xancus pyrum* extract. Similarly urea concentration in serum of cisplatin alone treated animals was increased, that is on 7th day 17.54 ± 0.4 mg/dL and on 11th day it was 19.71 ± 0.09 mg/dL which was significantly reduced to 15.01 ± 0.2 mg/dL on 7th day and 15.04 ± 0.12 mg/dL on 11th day by the administration of *Xancus pyrum* extract.

The urea level was increased drastically in serum of cisplatin alone treated animals, but this was significantly ($p < 0.001$) reversed by administration of *Xancus pyrum* extract.

Cisplatin treated animals showed an increase in the level of creatinine 1.438 ± 0.09 mg/dL on 7th day and 1.457 ± 0.08 mg/dL on 11th day in serum which was reversed to 1.07 ± 0.04 mg/dL on 7th day and 0.92 ± 0.08 mg/dL on 11th day by the administration of *Xancus pyrum* extract. It was also found that p-values was less than 0.001 ($p < 0.001$) showing that the cisplatin treated animals along with *Xancus pyrum* extract was statistically significant. Cancer is one of the dreadful diseases of this century. Radiotherapy and chemotherapy plays an important role in cancer treatment. Radiotherapy and chemotherapy is associated with toxic effect. They kill not only the tumour cell but also normal cells. Both these effects are associated with suppression of immune system. Most of the synthetic chemotherapeutic agents available today are immunosuppressant, cytotoxic and exert several side effects [17]. Modulation of immune system by cytotoxic agents is emerging as a major area in pharmacology, especially in case where undesired immunosuppression is the result of therapy. A major drawback of current cancer therapeutic practices such as chemotherapy and radiation therapy is bone marrow suppression

Group	Serum			
	Urea concentration (mg/dL)		BUN concentration (mg/dL)	
Days	7th day	11th day	7th day	11th day
Normal	28.41 ± 1.1	31.19 ± 1.4	13.26 ± 0.5	14.56 ± 0.6
Cisplatin alone	17.54 ± 0.4	4.71 ± 0.09	8.19 ± 0.20	2.20 ± 0.04
Cisplatin + *Xancus pyrum*	15.01 ± 0.2***	15.04 ± 0.12***	7.00 ± 0.12***	7.02 ± 0.05***

Values are expressed as mean ± SD.
$^{*}p < 0.05$.
$^{**}p < 0.01$.
$^{***}p < 0.001$.

Table 5.
Effect of Xancus pyrum treatment on the serum urea levels in cisplatin treated animals.

Group	Serum (mg/dL)	
Days	7th day	11th day
Normal	1.750 ± 0.04	1.880 ± 0.04
Cisplatin alone	1.438 ± 0.09	1.457 ± 0.08
Cisplatin + *Xancus pyrum*	1.075 ± 0.04***	0.920 ± 0.05***

Values are expressed as mean ± SD.
$^{*}p < 0.05$.
$^{**}p < 0.01$.
$^{***}p < 0.001$.

Table 6.
Effect of Xancus pyrum treatment on the serum, liver creatinine levels in cisplatin treated animals.

resulting in cytopenia and subsequent suppression of humural and cellular as well as nonspecific & specific cellular responses [18]. Weight of all relative organs was also increased in cisplatin treated animals by the extract administration, providing supportive evidence for *Xancus pyrum* extract immunostimulative potential during treatment of cisplatin. The effect of *Biophytum sensitivum* on the bone marrow cellularity and α-esterase positive cells after the administration of the methanolic extract of *Biophytum sensitivum* showed a significant ($p < 0.001$) enhancement in the bone marrow cellularity (28.3×10^6 cells/femur) compared to the normal control (17.3×10^6 cells/femur) animals. Moreover the number of α-esterase positive cells was also found to be increased significantly ($p < 0.001$) in the *Biophytum sensitivum* treated animals (1421 cells/4000 bone marrow cells) compared to the normal animals (905 cells/4000 bone marrow cells [19]. Similarly, the effect of *Xancus pyrum* on the bone marrow cellularity and α-esterase positive cells after the administration of the methanolic extract of *Xancus pyrum* showed a significant ($p < 0.001$) enhancement in the bone marrow cellularity in cisplatin treated animals, there was a drastic reduction in the number of bone marrow cells ($25.5 \times 10^5 \pm 1.414$ cells/femur) and α-esterase positive cells (634.5 ± 3.05 positive cells/4000 cells) compared to *Xancus pyrum* treated along with cisplatin animals. Treatment with *Xancus pyrum* could elevate the bone marrow cellularity and number of α-esterase positive cells. In cisplatin treated group of animals along with *Xancus pyrum*, bone marrow cellularity and α-esterase positive cells was found to be $69.7 \times 10^5 \pm 4.24$ cells/femur and 1227 ± 1.414 cells/4000 bone marrow cells. In the present study, chemoprotective effect of *Xancus pyrum* an important edible gastropod was found in mice. Administration of *Xancus pyrum* was found to increase the number of bone marrow cells significantly indicating the extract could stimulate the haematopoetic system. Moreover there was increased presence of α-esterase positive bone marrow cells indicating the extract treatment could also enhance the differentiation of stem cells. The extract was found to stimulate the weight of spleen and thymus indicating that *Xancus pyrum* stimulated the production of immune cells. The increased SGOT and SGPT levels in the serum of cisplatin treated mice can be attributed to the damaged structural integrity of the liver and kidney, because these enzymes are located in cytoplasm and are released into circulation after cellular damage [20]. The present study showed that *Xancus pyrum* extract had decreased the SGOT and SGPT levels in the serum during the cisplatin treatment in mice. The blood urea nitrogen (BUN) test is a measure of the amount of nitrogen in the blood in the form of urea, and a measurement of renal function. Urea is a substance secreted by the liver, and removed from the blood by the kidneys. The most common cause of an elevated BUN is poor kidney function. A greatly elevated BUN (>60 mg/dL) generally indicates a moderate-to-severe degree of renal failure. Impaired renal excretion of urea may be due to temporary conditions such as dehydration or shock, or may be due to either acute or chronic disease of the kidneys themselves [21]. Cisplatin administration in mice was found to increase the BUN concentration in serum and liver on 7th day and 11th day but this was significantly reduced to by the administration of *Xancus pyrum* extract, thus it's clear that the poor kidney function was enhanced by the *Xancus pyrum* extract. Cyclophosphamide treated animals showed an increase in the level of creatinine 1.536 ± 0.0603 mg/dL on 11th day and 1.526 ± 0.03 mg/dL on 15th day in serum which was reversed to 1.17 0.08 mg/dL on 11th day 0.87 mg/dL on 15th day by the administration of *Bauhinia tomentosa* [22].

Similarly, cisplatin administration in mice was found to increase the level of creatinine 1.438 ± 0.09 mg/dL on 7th day and 1.457 ± 0.08 mg/dL on 11th day in serum which was reversed to 1.07 ± 0.04 mg/dL on 7th day and 0.92 ± 0.08 mg/dL on 11th day by the administration of *Xancus pyrum* extract. Many well-recognized problems are associated with excessive intake of dietary fat, including obesity, insulin resistance, coronary heart disease, and some forms of cancer. While intakes of saturated, trans, and arachidonic fatty acids have been linked to the development of chronic disease, research shows omega-3 (n-3) fatty acids, specifically fish oils or in marine mollusks are essential in the prevention and treatment of disease. It is scientifically proven marine gastropod—*Xancus pyrum* is an edible gastropod which contains 8–10% of protein, 4–5% of carbohydrates 2–3% of minerals, and 1–2% of fat. This also contains omega 3 fatty acids [23]. The cisplatin treated animals showed high reduction in the weight of all the organs such as 0.23 ± 0.02 g/100 g body weight of spleen, 0.17 ± 0.01 g/100 g body weight of thymus, 3.7 ± 0.19 g/100 g body weight of liver, 1.2 ± 0.01 g/100 g body weight of kidney, 0.62 ± 0.01 g/100 g body weight of lungs. The cisplatin treated along with *Xancus pyrum* mice showed a significantly increase in the weight of all the organs such as 0.34 ± 0.072 g/100 g body weight of spleen, 0.23 ± 0.01 g/100 g body weight of thymus, 4.84 ± 0.05 g/100 g body weight of liver, 1.37 ± 0.18 g/100 g body weight of kidney, 0.769 ± 0.05 g/100 g body weight of lungs.

The p-values of cisplatin treated animals along with *Xancus pyrum* for spleen was $p < 0.01$, which was less significant but for thymus, liver, kidney and lungs it was $p < 0.05$ which was considered to be statistically significant. The number of bone marrow cells as well as α-esterase positive cells was decreased drastically in cisplatin alone treated animals, but this was significantly ($p < 0.001$) reversed by administration of *Xancus pyrum*. In cisplatin treated animals, on the 7th day there was a drastic reduction in the number of bone marrow cells ($25.5 \times 10^5 \pm 1.414$ cells/femur) and α-esterase positive cells (634.5 ± 3.05 positive cells/4000 cells) compared to *Xancus pyrum* treated along with cisplatin animals. Treatment with *Xancus pyrum* could elevate the bone marrow cellularity and number of α-esterase positive cells. In cisplatin treated group of animals along with *Xancus pyrum*, bone marrow cellularity and α-esterase positive cells was found to be $68.30 \times 10^5 \pm 4.24$ cells/femur and 1179 ± 2.121 cells/4000 bone marrow cells respectively on 7th day and it was again enhanced to $69.7 \times 10^5 \pm 4.24$ cells/femur and 1227 ± 1.414 cells/4000 bone marrow cells on 11th day respectively compare to the cisplatin alone treated animals ($20.93 \times 10^5 \pm 3.055$ cells/femur and 620.66 ± 3.05 cells/4000 bone marrow cells).

A significant increase of the levels of SGOT (82.280 ± 2.7 IU/L) and SGPT (85.22 ± 2.393 IU/L) observed in the serum samples of cisplatin alone treated group was reversed by the administration of *Xancus pyrum*. Treatment with cisplatin along with *Xancus pyrum* significantly reduced the levels of SGOT (52.68 ± 0.46 IU/L) and SGPT (55.820 ± 1.814 IU/L) in serum. Cisplatin treated animals showed a decrease in the levels of SGOT (32.67 ± 2.7 IU/L) & SGPT (42.04 ± 1.9 IU/L) observed in the liver sample. Administration of *Xancus pyrum* significantly increased the level of SGOT (41.545 ± 1.3 IU/L) and SGPT (48.290 ± 1.4 IU/L) in liver. It was also found that p-values was less than 0.001 ($p < 0.001$) showing that the cisplatin treated animals along with *Xancus pyrum* extract was statistically significant.

The renal functions can be estimated by biochemical parameters like BUN (blood urea nitrogen) and creatinine. Cisplatin administration in mice was found to increase the BUN concentration in serum on 7th day 18.19 ± 0.2 mg/dL and 11th day 2.20 ± 0.04 mg/dL but this was significantly reduced to 7.00 ± 0.12 mg/dL on 7th day and 7.025 ± 0.05 mg/dL on 11th day by the administration of *Xancus pyrum* extract. Similarly urea concentration in serum of cisplatin alone treated animals was increased, that is on 7th day 17.54 ± 0.4 mg/dL and on 11th day it was 19.71 ± 0.09 mg/dL which was significantly reduced to 15.01 ± 0.2 mg/dL on 7th day and 15.04 ± 0.12 mg/dL on 11th day by the administration of *Xancus pyrum* extract. It was also found that p-values was less than 0.001 ($p < 0.001$) showing that the cisplatin treated animals along with *Xancus pyrum* extract was statistically significant. Cisplatin treated animals showed an increase in the level of creatinine 1.438 ± 0.09 mg/dL on 7th day and 1.457 ± 0.08 mg/dL on 11th day in serum which was reversed to 1.07 ± 0.04 mg/dL on 7th day and 0.92 ± 0.08 mg/dL on 11th day by the administration of *Xancus pyrum* extract. It was also found that p-values was less than 0.05 ($p < 0.05$) showing that the cisplatin treated animals along with *Xancus pyrum* extract was statistically significant.

In the past several decades, thousands of marine compounds with tremendous pharmacological activities have been isolated and more than a dozen of them are in different stages of human clinical trials against various diseases. Thus from above mentioned experiment it is clearly known that the gastropod *Xancus pyrum* has reduced the side effects that is been caused by cisplatin (chemo drug). In the present study, chemoprotective effect of *Xancus pyrum* an important edible gastropod was studied. Administration of *Xancus pyrum* was found to increase the number of bone marrow cells significantly indicating the extract could stimulate the haematopoetic system. Moreover there was increased presence of α-esterase positive bone marrow cells indicating the extract treatment could also enhance the differentiation of stem cells. The extract was found to stimulate the weight of spleen and thymus indicating that *Xancus pyrum* stimulated the production of immune cells.

5. Conclusion

Cisplatin and its metabolites can bind to DNA, causing damage that may result in chromosome breaks, micronucleus formation and cell death. Administration of *Xancus pyrum* extract in cisplatin treated mice, found to enhance the bone marrow cellularity and α-esterase positive cells, which were drastically reduced in cisplatin alone treated control animals suggests that cisplatin induced myelosuppression was reversed or inhibited by *Xancus pyrum* extract administration possibly through its chemoprotective activity.

Author details

Jayaprakash Bindhu and Das Arunava*
Department of Biotechnology, Molecular Diagnostics and Bacterial Pathogenomics Research Laboratory, Bannari Amman Institute of Technology, Sathyamangalam, India

*Address all correspondence to: arunavadas@bitsathy.ac.in

References

[1] Proksch P, Edrada RA, Ebel R. Drugs from the seas—Current status and microbiological implications. Application of Microbial Biotechnology. 2002:125-134

[2] Mayer AM, Gustafson KR. Marine pharmacology in antitumor and cytotoxic compounds. International Journal on Cancer. 2003:291-299

[3] Garcia-Fernandez LF, Reyes F, Sanchez-Puelles JM. The marine pharmacy, new antitumoral compounds from the sea. Pharmaceutical News. 2002;**9**:495-501

[4] Haefner B. Drugs from the deep marine. Drug Discovery Today. 2003;**8**: 536-544

[5] Goodman LS, Wintrobe MM, Dameshek W, Goodman MJ, Gilman A, McLennan MT. Nitrogen mustard therapy. The Lancet Oncology. 2002: 221-225

[6] Girdhani S, Bhosle SM, Thulsidas SA, Kumar A, Mishra KP. Potential of radio sensitizing agents in cancer chemo-radiotherapy. Journal on Cancer Therapeutics. 2005:125129

[7] Jaiprakash, Gupta SK. Natural products for chemoprevention. Indian Journal of Medical and Pediatric Oncology. 2004;**25**:3-37

[8] Bouchet P. New records and species of *Abyssochrysos* (Mollusca, Caenogastropoda). Journal of Natural History. 1991;**25**:305-313

[9] Venkatesan RM, Vijayakumaran, Gopal R, Kathiroli S. Marine organisms in Indian medicines and their future prospects. Natural Product Radiance. 2008;7(2):139-145

[10] Usmanghani K, Siddiqui M, Ahmad M, Kazmi MA. Composition of fatty acids of a gastropod, *Xancus pyrum*. Journal of Islamic Academy of Sciences. 1989:165-167

[11] Madhuri S, Pandey G. Some anticancer medicinal plants of foreign origin. Current Science. 2009;**96**:6-25

[12] Bruce NA, Lois SG, Walter CW. The causes and prevention of cancer. Proceedings of the National Academy of Sciences of the United States of America. 1995;**92**:5258-5265

[13] Gralla RJ, Nancy G, Houlihan. Understanding and managing chemotherapy side effects. In: Cancer Care. 2nd ed. 2008. pp. 1-16

[14] Dahanuhar SA, Kulkarni RA, Rege NN. Pharmacology of medicinal plants and natural products. Indian Journal on Pharmacology. 2000;**32**: 81-118

[15] Pallabi S, Somvanshi VS, Rao TA. Health of the coastal marine environment and fisheries development in India. In: Vision on Indian Fisheries of 21st Century. 1998. pp. 276-287

[16] Nicolau KC, Hepworth D, King NP, Finlay MRV. Chemistry biology and medicine of selected tubulin polymerizing agents. Pure and Applied Chemistry. 1999;**71**:989-997

[17] Chen P, Huang W-B, Wang K-J. Immunomodulation in the marine gastropod *Haliotis diversicolor* exposed to benzo(a)pyrene. Oceanography and Environmental Science. 2003:132-144

[18] Aditya SA. Synthesis of some phenacyl nicotinate as potential anti-cholesterolemic agent. Proceedings of the National Chemotherapeutic Conference II. 1976;**2**:46-59

[19] Moltedo S, Rashid S, Lodhi F, Ahmad M. Preliminary cardiovascular

activity evaluation of capparidisine, a spermidine alkaloid from *Capparis decidua*. Pakistan Journal of Pharmacology. 2006;**6**:71-108

[20] Harvey GB. Testing of drugs inhibiting the formation of gastric ulcers. Journal of Pharmacological and Toxicological Methods. 1992;**8**:33

[21] Raphael TJ, Kutten G. Immunomodulatory activity of naturally occurring monoterpenes carvone, limonene and perillic acid. Journal of Immunopharmacology and Immunotoxicology. 2003;**25**:285-294

[22] Sporn MB, Dunop NM, Newton DL, Smith JM. Prevention of chemical carcinogenesis by Vitamin A and its synthetic analogs (retinoids). Proceedings of the National Seminar on Cancer and Prevention. 1976;**35**: 1332-1338

[23] Biella CDA, Salvador MJ, Aparecida D. Evolution of immunomodulatory and anti-inflammatory effects and phytochemical screening. Memórias do Instituto Oswaldo Cruz, Rio de Janeiro. 2008;**103**(6):569-573

Chapter 4

Intermediate Host Snails of Human Schistosomes in the Senegal River Delta: Spatial Distribution According to Physicochemical Parameters

Raphael Abdoulaye Ndione, Sidy Bakhoum, Chistopher Haggerty, Nicolas Jouanard, Simon Senghor, Papa Demba Ndao, Gilles Riveau and Cheikh Tidiane Ba

Abstract

The objective of the study was to evaluate the influence of physicochemical parameters of water on the spatial distribution of snail intermediate hosts of human schistosomes in the Senegal River Delta. Eight water points in three endemic villages for schistosomiasis were selected for biweekly monitoring of snail numbers and physicochemical parameters of water at the beginning of the rainy season. The results show that the spatial distribution of snail populations is a function of certain parameters. The pH, the dissolved oxygen and its saturation, and the temperature have a positive influence on the *Bulinus* and *Biomphalaria*, while the conductivity, the speed of flow, and the salts (phosphates, salinity, and nitrates) seem to act negatively on the populations of these snails.

Keywords: physicochemical parameters of water, snails, intermediate hosts, schistosomiasis, spatial distribution, Delta, Senegal

1. Introduction

Schistosomiasis is among the most widespread human parasitic diseases with more than 200 million people infected worldwide, with the majority of these infections occurring in sub-Saharan Africa. The human schistosomiasis species present in Senegal are *Schistosoma haematobium* and *S. mansoni* which are transmitted by contact with freshwater snails as an intermediate host and caused urinary or intestinal schistosomiasis, respectively. The intermediate host for *S. mansoni* belongs to the genus *Biomphalaria pfeifferi*, while genus *Bulinus* harbors *S. haematobium* [1–3]. *Biomphalaria pfeifferi* is the only species involved in the transmission of *S. mansoni* in Senegal [4]. It was in the late 1980s that the first cases of intestinal schistosomiasis were diagnosed in the town of Richard-Toll, north-east of the lower Senegal River valley [5]. The transformation of the environment by human agricultural activities favors the creation of breeding sites for the development of snails that spread to other sites [6, 7]. This situation has been aggravated by hydraulic developments, the construction of numerous small and large dams, as well as the multiplication of irrigation canals [8]. The modification of the practices of the populations bordering these developments is also concretized by intensification of human contact with infected water [9]. These factors contribute to the evolution of the incidence of these schistosome infections and their pathologies in the region [10].

In the present study, the importance of the physicochemical parameters of surface waters (pH, dissolved oxygen, conductivity, phosphates, salinity, nitrates, temperature, and flow velocity of water) on the spatial distribution of snail intermediate hosts of schistosomes in human beings has been studied.

2. Material and method

2.1 Study sites

The choice of prospecting sites was guided by their human and animal associations and the presence of human schistosomiasis transmission. The selected villages are bordered by creeks and tributaries of the river in which local people carry out domestic or work activities related to water (**Figure 1**).

In this study, the three villages that have been selected were co-exposed to *S. haematobium* and *S. mansoni* infection. The village of Menguègne Boye (ME) (16°017,315 N, −16°356,659 O) has four (04) water points (ME1, ME2, ME3, ME4), the village of Ndellé Boye (NE) (16°168,725 N, −16°289,124 O) has two (02) water points (NE1, NE2), and the village of Thilla (TA) (16°054,994 N, −16°331,894 O) also has two (02) water points (TA1 and TA2). NE1 and NE2 are separated by a dam and communicate with each other through transverse pipes.

The study covers the period from 03 July to 02 August 2017 (beginning of the rainy season), due to a survey every 15 days from 9 am to 12 am. Three surveys

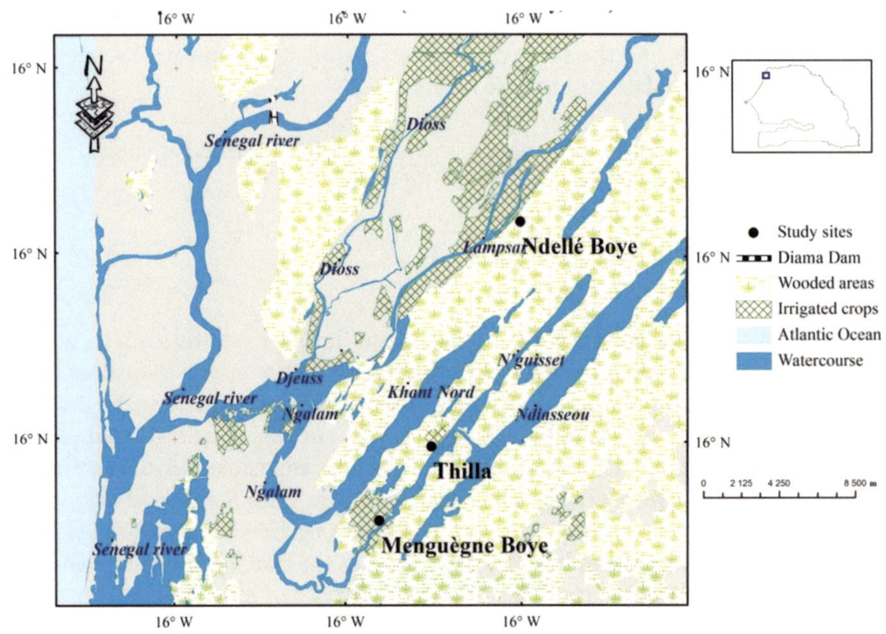

Figure 1.
Map of the study sites.

at the different water points constituting zones of human-water contact, zones of transmission of the disease, were carried out. In parallel with the malacological investigations, in situ measurements of the physicochemical factors were conducted.

2.2 Evaluation of physicochemical parameters

The evaluation of the physicochemical parameters was carried out to a depth not exceeding 0.5 meters. Conductivity (μS/cm), salinity (psu), dissolved oxygen (mg/l) and its saturation (%), and nitrate concentration (mg/l) were measured with the YSI 600 multiparameter digital probe recorder (HANNA instruments). The plunging probes were covered by a scraper. Phosphate measurement (mg/l) was performed by using the Phosphate High Range (PHR) model HI 96717 C (HANNA instruments). Briefly, 10 ml of water was first mixed in a tube with the reagent Phosphate HR Reagent B (HANNA instruments). Then the tube was placed in the device for 3 minutes to obtain the value. Hydric potential (pH) was obtained using an ESEE pH-meter (HANNA instruments) by immersion of a probe. The flow velocity (m/s) and the water temperature (°C) were measured by a flowatch (JDC electronic) equipped with a submerged propeller at a depth of 0.5 meters.

2.3 Malacological study

2.3.1 Snail collection

The presence of snails was looked for at each water point using a wire wick (2.5 mm) by diversifying the areas surveyed. The areas of prospect were the *Typha* area, floating vegetation, and mud. Depth was measured before each scoop with the dip net. The number of scoops was a function of the surface of the water point and varied between 10 and 15. After washing the vegetation in the basin, the water was filtered using a metal screen (2.5 mm) only passing water and fine debris. The snails were harvested with tongs. All snails from the same waterhole were grouped in one or more pots if necessary and brought to the laboratory for identification.

2.3.2 Identification of snails

The identification is based on the Mandahl-Barth key based on the morphology of the shell. Snails not identifiable to the eye were observed with a binocular magnifying glass. The latter method was mainly of interest to snails of the genus *Bulinus* because the species of the genus *Biomphalaria pfeifferi* is easily recognizable by the discoid shape of its shell. The density (d) of snails was expressed in (average) numbers of snails per scoop [11].

2.4 Statistical analyses

The R Studio, Excel software, and XLSTAT extension were used for the analysis of the results. Results are presented +/− SD averages. In order to establish a relationship between the different physical (conductivity, flow rate, water temperature) and chemical (dissolved oxygen, saturation, water salinity, pH, nitrate contents, and phosphates) parameters and the density of snails, a statistical principal component analysis (PCA) was applied to all variables. With XLSAT, the realization of principal component analysis allowed to analyze a table of observations/quantitative variables or a correlation or covariance matrix.

3. Results

3.1 Physicochemical parameters of the water points

The averages of the eight [8] abiotic factors measured in our eight water points are shown in **Figure 2**. The highest water temperature was observed at Mbenguègne Boye (ME3) with 29.9°C. Thilla with its first site (TA1) recorded 29.8°C, while Ndellé Boye had the lowest temperature from its second NE2 water point with 28.5°C. Only the water points NE2 and TA2 recorded a speed greater than 0 (zero) with, respectively, 1 m/s and 3 m/s. The maximum conductivity was obtained at Ndellé Boye: 200.26 µS/cm for NE1 and 195.16 µS/cm for NE2. The lowest conductivity content was found at the first Menguègne Boye point (ME1, 132.06 µS/cm). Dissolved oxygen (mg/l) showed +/− significant variations from 38.3 mg/l to ME3 to 5.7 mg/l to TA2. At Menguègne Boye the maximum dissolved O_2 content was observed at ME1

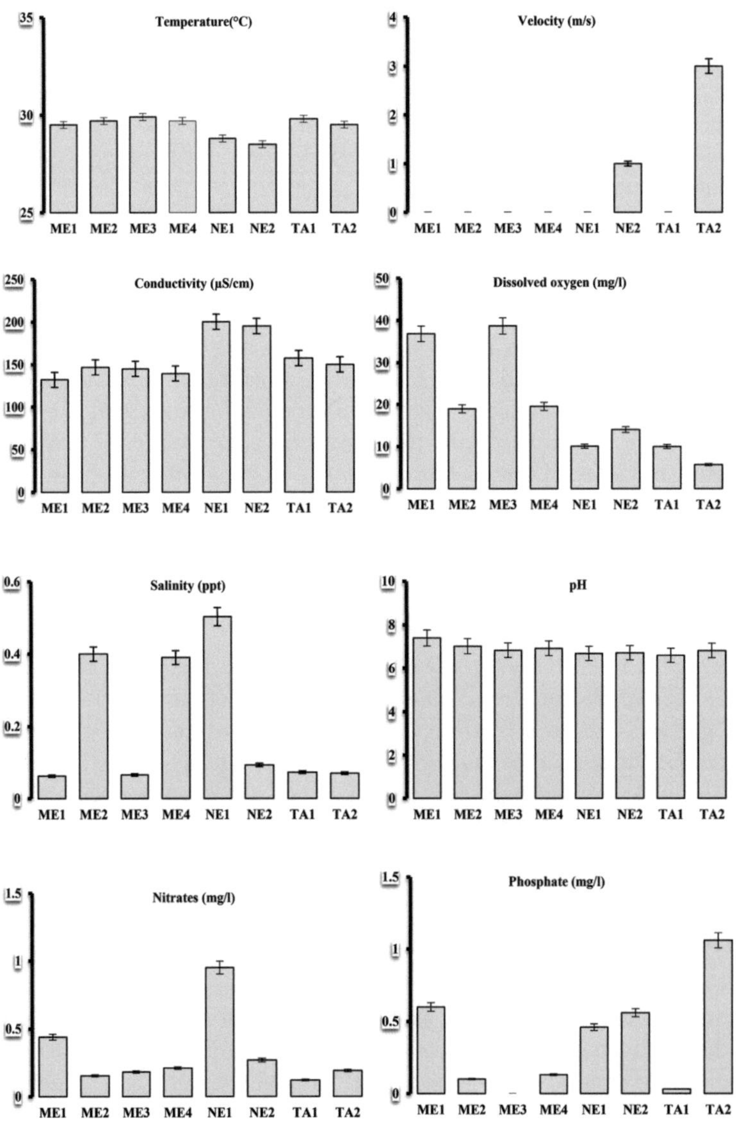

Figure 2.
Physicochemical data (average) of the water points.

Site	Water point	Biom.	B. fors.	B. glob.	B. sene.	B. trun.
Menguègne Boye	ME1	2.16 (± 0.64)	1.04 (± 0.31)	0.5 (± 0.15)	0.11 (± 0.03)	27.54 (± 8.26)
	ME2	0.3 (± 0.09)	1.1 (± 0.33)	0.46 (± 0.13)	0.93 (± 0.27)	4.43 (± 1.32)
	ME3	0	0	0,13 (± 0,04)	0	2,86 (± 0,85)
	ME4	0.06 (± 0.01)	0.2 (± 0.06)	0.13 (± 0.041)	0	1.26 (± 0.37)
Ndellé Boye	NE1	0	0	0	0	0,69 (± 0,20)
	NE2	0	0.47 (± 0.14)	0	0.38 (± 0.11)	5.32 (± 1.59)
Thilla	TA1	2.38 (± 0.71)	0.05 (± 0.01)	0.076 (± 0.02)	0	1.38 (± 0.41)
	TA2	0	0.06 (± 0.01)	0	0	0.06 (± 0.01)

Biom. = Biomphalaria, B. = Bulinus, fors. = forskalii, glob. = globosus, sene. = senegalensis, trun. = truncatus.
***ME**= Mbenguègne Boye, **NE**= Ndellé Boye, **TA**= Thilla*

Table 1.
Density of intermediate host snail of human schistosomes (number of molluscs/scoop).

36.83 mg/l and ME3 (38.7 mg/l). Thilla had the lowest dissolved O_2 content at point 2 (5.7 mg/l at TA2). The salt content obtained is very weak (1 ppt = 1–9 mg/l). We found in Ndellé (NE1) the largest salt measure (0.503 psu), followed by ME2 with 0.4 ppt. The lowest levels were found at ME1 (0.063 psu) and ME3 (0.066 psu). The pH showed its maximum values in Menguègne Boye—7.4 to ME1, 7.02 to ME2, and 6.92 at ME4—while Thilla recorded the lowest value at its first point (6.59). At Ndellé, we had the maximum nitrate content (0.95 mg/l) and the average phosphate levels: 0.46 mg/l at point 1 (NE1) and 0.56 mg/l at NE2. TA2 had the highest phosphate content (1.06 mg/l), and TA1 showed only 0.03 mg/l, while we found 0 mg/l at the third point of Menguégne Boye.

3.2 Malacological data

The malacofauna intermediate host of human schistosomiasis consisted of *Biomphalaria pfeifferi*, *Bulinus forskalii*, *B. globosus*, *B. senegalensis*, and *B. truncatus*. The total population of snails recovered in the three villages (eight water points) was 2068 snails. The analysis of the malacological data showed that *B. truncatus* colonized all study sites (**Table 1**). Its dominance within this stand was observed in all water points. However, ME1 had the highest density of *B. truncatus*, while *B. globosus* (0.5 individuals/scoop) is absent at Ndellé (NE1, NE2) and TA1. *B. forskalii* was mainly present in Menguègne Boye with d = 1.04 at ME1 and d = 1.1 at ME2. *B. senegalensis* was found only at Mbenguègne Boye (ME1 and ME2) and at the second point of Ndellé (NE2, d = 0.38). *Biomphalaria pfeifferi* was the second most represented species (d = 2.16 at ME1 and d = 2.38 at TA1) after *B. truncatus*.

3.3 Relationship between intermediate hosts snails of schistosomes and measured abiotic factors of the biotope

Figure 3 shows that the presence of intermediate snail hosts was positively correlated with temperature, pH and dissolved oxygen, and its saturation. Salts (phosphate, salinity, and nitrates), conductivity, and velocity did not have any direct effect on the presence of snails. The PCA indicates that the pH, the dissolved O_2, and its saturation (%) were strongly linked to the presence of the species *B. globosus, B. truncatus*, and *B. forskalii*, whereas they were moderately related to that of *Biomphalaria* as shown in the correlation matrix. Projections on PCA of salinity, nitrate, and phosphate levels of water are orthogonal to those of snails in general, indicating that there were no significant direct influences of these

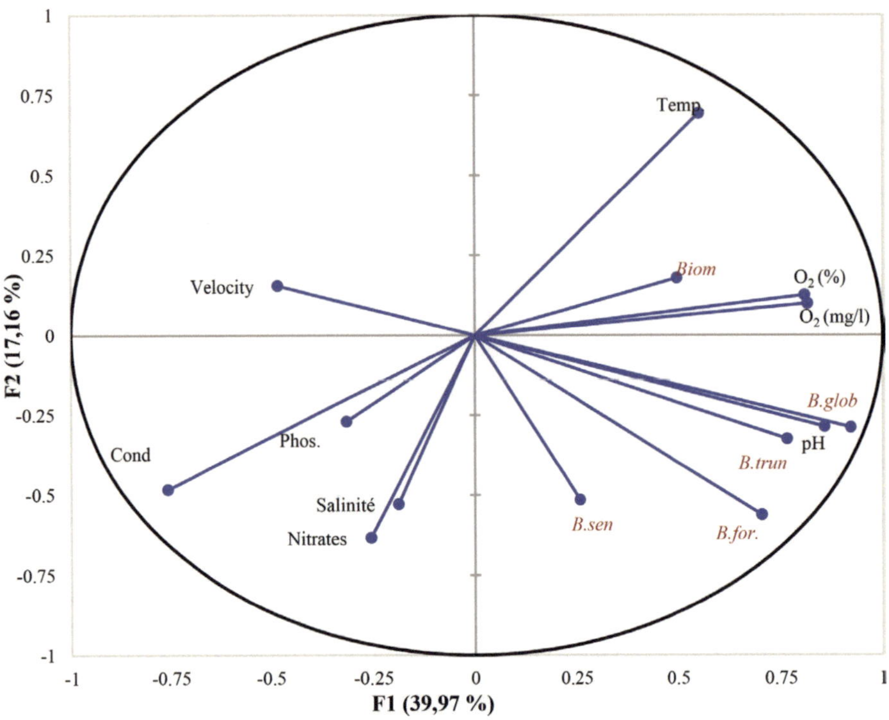

Figure 3.
Relationship between the density of intermediate host snails and the physicochemical parameters of water.

parameters on snail populations. The temperature is positively correlated with *Biomphalaria*. Conductivity, velocity, and salinity are more or less opposed to those of *B. globosus, B. truncatus*, and *B. forskalii*.

4. Discussion

The temperatures measured ranged from a minimum of 28.5°C to Ndellé Boye to a maximum of 29.9°C to ME3. The lowest temperatures recorded at Ndellé could be explained by the fact that the freshwater ecosystem found in this environment is less exposed to light because of its highly developed eutrophication. It has been shown [11] that during the month of July, water temperatures on the other side of the Senegal River (Right River) vary between 26.1°C and 28.9°C. pH values vary almost by one unit (between 6.59 and 7.4). Its slight basicity accompanies the words of N'Diaye *et al*. [12] who said that the pH values measured in the waters of the Senegal River place the latter in the excellent to good level of surface water ($6.5 < pH < 8.5$). Phosphate and nitrate contents would not only come from agricultural land (leaching of fertilizers by rainwater) but also from household activities for which women use different detergents and antiseptics. Halstead et al. [13] shows the potential impact of agrochemicals on the transmission of schistosomiasis and therefore on snails. In a recent study in the same region, Diallo et al. [14] argue that washing clothes directly in the river alone probably justifies the presence of phosphates. They also argue that rains can be an important vector for the transport of domestic waste, urine, and excrement of animals to the river by the phenomenon of leaching. The strong eutrophication noted on the biotope particularly to Ndellé Boye would come from the important contents of nitrates and phosphates in the water. In addition, the presence of the Diama Dam and the association of dikes reduce the oxidation and flow velocity of the valley water [15]. These facts, combined with waste dumps around the villages and cities and agrochemicals from irrigated fields, accelerated eutrophication of water in the Senegal River Valley. The very low salt levels indicate that the watercourses of our study sites are freshwater ecosystems. The electrical conductivity which is a measure of the ability of an aqueous solution to conduct electrical current shows significant variations (132.06 and 200.26 μS/cm). The high levels obtained to Ndellé (200, 26 μS/cm at NE1 and 195.16 μS/cm to NE2) are due to the strong contributions of organic matter in the water, resulting in their greater mineralization. By comparing the values of the conductivity measured at the WHO reference water level, which is 300 μs/cm [16], we deduce that the water from these effluents of the Senegal River is of good quality. However, a study by Tfeila et al. [17] on the Senegal River indicates a much smaller variation (47.4 and 67.1 μS/cm) than that observed in our study sites.

The simultaneous presence of both intermediate host snail species is indicative of the existence of both schistosomiases in the study area. Diaw et al. [18] note the presence of *Biomphalaria pfeifferi, Bulinus globosus, B. truncatus, B. senegalensis*, and *B. forskalii* at Mbodiène bordered by Lampsar, which was a major focus of schistosomiasis. It also shows the existence of these organisms in the delta with an increase in populations of *Biomphalaria* and a wider distribution. Five [5] species of snail intermediate hosts of human schistosomes in Senegal were found in our study sites. *B. senegalensis*, an intermediate host of *S. haematobium*, is very common and abundant in the regions of Saint-Louis, Tambacounda, Kaolack, and Fatick [19, 20]. We thus find its presence in Menguègne Boye (d = 0.11 for ME1 and 0.93 for ME2). *B. globosus* and *B. truncatus*, intermediate hosts of *S. haematobium*, are very commonly found in the Senegal River delta. *Biomphalaria pfeifferi*, the main intermediate host of *S. mansoni*, was not found in Ndellé Boye although its presence has been previously mentioned [21] in Lampsar. These authors indicate that in the delta, this species was the most abundant in the early 1990s. However, the presence of this snail in Thilla and Menguègne Boye with respective densities of 2.38 and 2.16 was quite poor, which has been confirmed by the work of Ndir [20] who supports a reduced presence in the delta but with a range tending to extend toward the southern region (Louga region) since the impoundment of the dams. The dominance of *B. truncatus* in our study sites is very remarkable. This could be explained by the fact that the sites Menguègne Boye, Ndellé Boye, and Thilla are permanent watercourses. Gbocho et al. [16] confirm that they are favorable to the proliferation of this species.

The presence of biotopes favorable or not to the life of the snails is due to ecological transformations of the environments. Among these transformations, we can note the presence of vegetation that could determine the presence or absence of snails [22]. The physicochemical conditions that accompanied ecological changes could influence the distribution of snails. A high rate of water conduction has been noted at TA2; the rarity of snails observed at this point of water could be due to speed. Speed is a physical factor that opposes the residence of snails if it exceeds 0.3 m/s [19]. The values obtained for oxygen and pH are favorable for the habitat of gastropods. The study shows a positive correlation between temperature and *Biomphalaria*, whereas temperatures of 29.9°C have no influence on the *Bulinus*. In a study done in the lake of the Taabo Dam in Côte d'Ivoire, Gbocho et al. [16] show a positive correlation of temperature with *Biomphalaria* at a maximum T° of 31.5°, whereas this temperature acts negatively on the populations of *Bulinus*. The measured salinity has no significant influence on snails (except *Biomphalaria* and *B. truncatus* species). This could be due to its very low content. Diaw et al. [19] give the example of the effects of this parameter on the development of snails in the delta of the Senegal River where these gastropods have proliferated after the start of operation of the Diama anti-salt dam. Significant values of conductivity obtained at Ndellé

(200, 26 μS/cm at NE1 and 195.16 μS/cm to NE2) generally have a negative effect on the density of snails. Which could confirm the remarks of Gbocho et al. [16] who argue that conductivity levels (74 and 77.4 μS/cm) that they observed do not correlate significantly with intermediate host populations because of its low values. Our study demonstrates that physicochemical parameters such as conductivity, dissolved oxygen and its saturation, pH, and water flow velocity have an important role on the spatial distribution of snail intermediate hosts of human schistosomiasis.

5. Conclusion

The study of the influence of physicochemical parameters on the spatial distribution of intermediate snail hosts in human schistosomes in the Senegal River delta provided insights on the diversity of snail and their density and, secondly, the role of these parameters in the distribution of snails. The presence of snail intermediate hosts of human schistosomes would be conditioned by the temperature, oxygen, and pH with which they are positively correlated. An increase in salt (phosphates, nitrates, and salinity), conductivity, and velocity would lead to a lower density of snails.

Author details

Raphael Abdoulaye Ndione[1,2*], Sidy Bakhoum[2], Chistopher Haggerty[3], Nicolas Jouanard[1], Simon Senghor[1], Papa Demba Ndao[1], Gilles Riveau[1] and Cheikh Tidiane Ba[2]

1 Biomedical Research Center EPLS, Saint Louis, Senegal

2 Cheikh Anta DIOP University, Dakar, Senegal

3 Department of Integrative Biology, University of South Florida, Tampa, FL, USA

*Address all correspondence to: raphabdou@gmail.com

References

[1] Isabwe A, Ruberanziza E, Mupfasoni D, Ruxin J, Clerinx J, White PT. Potential for transmission of schistosomiasis in Kayonza. Rwanda Medical Journal. 2012;**6**:8

[2] Sturrock RF. The parasites and their life cycles. Human Schistosomiasis Rwanda Medical Journal. 1993;**69**:14-19

[3] Sy I, Diawara L, Ngabo D, Barbier D, Dreyfuss G, Georges P. Bilharzioses au Sénégal oriental prévalence chez les enfants de la région de Bandafassi. La Medicina Tropical. 2008;**68**:267

[4] Campbell G, Noble LR, Rollinson D, Southgate VR, Webster JP, Jones CS. Low genetic diversity in a snail intermediate host (*Biomphalaria pfeifferi* Kross, 1848) and schistosomiasis transmission in the Senegal River basin. Molecular Ecology. 2010;**19**:241-256

[5] Talla I, Kongs A, Verle P, Belot J, Sarr S, Coll AM. Outbreak of intestinal schistosomiasis in the Senegal River basin. Annales de la Société Belge de Médecine Tropicale. 1990;**70**:173-180

[6] Symoens JJ, Burgis M, Gaudet JJ, others. Ecologie et Utilisation des Eaux Continentales Africaines. UNEP; 1981

[7] Steinmann P, Keiser J, Bos R, Tanner M, Utzinger J. Schistosomiasis and water resources development: Systematic review, meta-analysis, and estimates of people at risk. The Lancet Infectious Diseases. 2006;**6**:411-425

[8] Traoré M. Importance des aménagements hydrauliques dans la transmission des schistosomoses. In: Chippaux J.P. Lutte Contre les Schistosomes en Afrique de l'Ouest; 15-18 Février 2000; Niamey-Cermes. Paris: éditions IRD. 2000. p. 292

[9] Helmut K, Andréa G, others. Microgeographical patterns of schistosomiasis and water contact behavior; examples from Africa and Brazil. 1998. pp. 37-50

[10] Cecchi P, Baldé S, Yapi YG. Mollusques hôtes intermédiaires de bilharzioses dans les petits barrages. L'eau en partage. Les petits barrages de Côte d'Ivoire. 2007. pp. 175-189

[11] Mandahl-Barth G. Les hôtes intermédiaires de *Schistosoma*: *Biomphalaria* et *Bulinus* africains. *Danemarks Akvarium, Charlottenlund, Danemark,* Palais des Nations, Genève, 16, pp. 1103-1163 et 17, pp. 1-65. 1959

[12] N'Diaye AD, Salem KMM. Contribution a l'étude de la qualité physico-chimique de l'eau de la rive droite du fleuve Sénégal. Larhyss Journal. Janvier 2013;**12**:71-83. ISSN: 1112-3680

[13] Halstead NT, Hoover CM, Arakala A, Civitello DJ, De Leo GA, Gambhir M, et al. Agrochemicals increase risk of human schistosomiasis by supporting higher densities of intermediate hosts. Nature Communications. 2018;**9**:837. DOI: 10.1038/s41467-018-03189

[14] Diallo AD, Namr KI, N'Diaye AD, Garmes H, Kankou M, Wane O. L'intérêt des méthodes d'analyses statistiques dans la gestion du suivi de la qualité physico-chimique de l'eau de la rive droite du fleuve Sénégal. Larhyss Journal. 2014;**17**:101-114. ISSN: 1112-3680

[15] Manikowski S, Strapasson A. Sustainability assessment of large irrigation dams in Senegal: A cost-benefit analysis for the Senegal River valley. Frontiers in Environmental Science. 2016;**4**:18

[16] Gbocho FY, Diakité RN, Akotto FO, N'Goran EK. Dynamique des populations de mollusques hôtes

intermédiaires de *Schistosoma haematobium* et *Schistosoma mansoni* dans le lac du barrage de Taabo (sud Côte d'Ivoire). Journal of Animal & Plant Sciences. 2015;**25**:3939-3953

[17] Tfeila MM, MOSA KO, Souabi S, Aboulhassan MA, Taleb A, Bouezmarni M. Suivi de la qualité physicochimique de l'eau du fleuve Sénégal: Cas du captage du Beni Nadji alimentant en eau potable les wilayas de Nouakchott (monitoring of water physico-chemical quality of the Senegal River: The case of capture of Beni Nadji supplying drinking water of the Wilaya of Nouakchott). Journal of Materials and Environmental Science. 2016;**7**(1):148-160

[18] Diaw OT, Vassiliades G, Seye M, Sarr Y. Rôle épidémiologie des mollusques du genre Bulinus dans la transmission des schistosomiases animales et humaines au Sénégal. Revue Sénégalaise des Recherches Agricoles et Halieutiques. 1988;**1**(1):74-78

[19] Diaw OT, Ndir O, Toupane MG. Guide de surveillance malacologique et de lutte contre les mollusques hôtes intermédiaires des bilharzioses. Ministère de la santé: Service national des grandes endémies; 1999. p. 60

[20] Ndir O. Situation des schistosomes au Sénégal. Lutte contre les schistosomes en Afrique de l'Ouest. Paris: Editeur scientifique; 2000. pp. 225-236

[21] Ernould JC. Epidémiologie des schistosomoses humaines dans le delta du fleuve Sénégal. Phénomène récent de compétition entre *Schistosoma haematobium* Sambon, 1907 et *S. mansoni* (Bilharz, 1852) [thèse de doctorat]. Université Paris XII; 1996. p. 589

[22] Ndione RA. Etude de l'influence des paramètres environnementaux sur la dynamique des populations de mollusques hôtes intermédiaires des schistosomes humains dans la vallée du fleuve Sénégal (cas de la commune de Richard-Toll). Mémoire de Master 2 en Biologie Animale: Ecologie et gestion des écosystèmes. Université Cheikh Anta Diop de Dakar; 2015. p. 30

Section 2

Management

Chapter 5

Reproductive Biology, Seed Production, and Culture of the Hawaiian Limpet *Cellana sandwicensis* (Pease, 1861)

Hua Thai Nhan and Harry Ako

Abstract

The purpose of this chapter was to describe the current finding on the development of aquaculture technologies for the Hawaiian limpet *Cellana sandwicensis*, known as "yellow opihi" in Hawaii. Some reproductive biology characteristics of *C. sandwicensis* were reported including spawning season, gonad maturation stages, maturity size, and fecundity. Monthly record of gonadosomatic index (GSI) suggested that the natural spawning season of *C. sandwicensis* occurred from November to January. Attempting studies on seed production have also performed and achieved several important key points such as inducing final maturation by incorporating arachidonic acid (ARA) into the diet and injecting salmon gonadotropin-releasing hormone analog (sGnRHa). Laval metamorphosis and settlement were successfully induced using a combination of algae *Palova* and benthic diatom *Amphora*. Stomach content analysis gave an insight into the palatability factor for further development of artificial feed; later on, the algae *Porphyra* commonly known as Nori was as attractive as a biofilm and was used as a feeding stimulant. Nutritional study on specific nutrient requirements such as protein, carbohydrate, and energy has been conducted and found that dietary 35% protein, 32% carbohydrate, and protein to energy (PE) ratio ranging from 87.2 to 102.9 mg/kcal could be used for the development of commercial feed for limpet *C. sandwicensis*.

Keywords: Hawaiian limpet *Cellana sandwicensis*, yellow opihi, spawning season, seed production, nutrient requirement

1. Introduction

Limpets are marine gastropods. They distribute at different intertidal zones of most oceans, from the upper littoral to the shallow subtidal on the rocky coasts. They feed by grazing on macroalgae, benthic diatom growing on rocky substrate because they attach themselves to rocks, and/or any substratum using pedal mucus and a muscular "foot," which also enables them to go against dangerous wave action, desiccation, and predator.

Cellana genus is a marine gastropod mollusk in the family Nacellidae [1]. This genus distributes in the temperate and tropical Indo-Pacific Oceans, Hawaii,

Australia, and New Zealand. Species are also found around the coasts of Japan, the Red Sea, Madagascar, South Africa, and the subantarctic island. There are more than 58 species of this genus. Among of those, many of them are of high economic value and aquaculture, for example, the two species *Cellana talcosa* and *Cellana sandwicensis* are expensive in Hawaii.

In Hawaii, there are three main endemic Hawaiian limpet species, called "opihi," including black foot or makaiauli (*Cellana exarata*), yellow foot or ālinalina (*C. sandwicensis*), and the largest species, giant limpet or kō'ele (*C. talcosa*) [2]. Natural ecological distribution of Hawaiian limpet is different intertidal zones of habitation on rocky shores. *C. exarata* is commonly found at higher intertidal zones, and *C. sandwicensis* is at low intertidal zones and rarely exposed by tide, whereas *C. talcosa* distributes in deep water [3–6]. These species are herbivorous grazers that feed on benthic microalgae, diatoms that growing hard substrates such as rocky substrates, death coral reef, and so on. They use teeth in their radula to graze on the toughest crustose coralline algae [3]. They are considered as high-value food market and high-potential candidate species for commercial aquaculture. High commercial catch reduced significantly from 150,000 pounds in the 1900s to about 10,000 pounds in 1978 [7]. The scarcity has boost up prices to about $200/gallon with shell on [8].

In addition to important food sources, these Hawaiian limpet species are also culturally important in Hawaiian society. Many people (opihi pickers) were asked to collect these opihi for parties or family gathering with high prices. Besides that limpet's shell also continue to be used as tools for scraping skin off taro plant and sweet potato and grating coconut meat before eating [9] and as decorative elements in jewelry.

The success of any aquaculture species depends on seed production in captivity. Understanding the completion of the life cycle of limpet would make limpet aquaculture sustainable. The first priority is to understand some reproductive characteristics of the limpet species. This would also provide us with better knowledge for breeding limpet in the hatchery. In this chapter, we describe the current finding on reproductive biology, seed production, nutrient requirement, and culture techniques for the Hawaiian limpet *Cellana sandwicensis*.

2. Some reproductive characteristics of Hawaiian limpet *C. sandwicensis*

Reproductive characteristics of Hawaiian limpet *C. sandwicensis* have been reported by several studies [10–12]. The main focused reproductive criteria were spawning season, gonad development stage, fecundity, and maturity size.

2.1 Spawning season

A total of 266 specimens (**Table 1**) were sampled for a 1-year cycle (November 2011–December 2012) in Hawaii Island, and gonadal somatic index (GSI) was determined. The GSI was calculated according to equation GSI = (GW/BW) × 100, where GW is gonad weight and BW is body weight or soft body tissue. Gonad development stage was also evaluated by using histological examination. The result showed that the highest average GSI of *C. sandwicensis* was noticed between November and January. This suggests that spawning seasons of *C. sandwicensis* may occur right after this period, whereas the lowest GSI was found from March to August, this could probably be the resting season of the species. Similarly, the same GSI pattern (**Figure 1**) of males and females *C. sandwicensis* suggested synchronized spawning of male and female *C. sandwicensis* in the wild [10].

Date of sampling	n	Shell length (cm)	Total weight (g)	Body weight (g)	Gonad weight (g)	GSI (%)
November 12, 2011	13	4.26 ± 0.59	15.3 ± 5.49	5.93 ± 2.43	1.68 ± 0.92	26.8 ± 6.27
December 04, 2011	30	3.46 ± 0.51	7.67 ± 3.63	2.54 ± 1.39	0.31 ± 0.22	11.7 ± 5.22
January 31, 2012	17	3.55 ± 0.96	9.31 ± 5.74	2.97 ± 2.40	0.93 ± 1.06	22.9 ± 12.4
February 28, 2012	27	3.11 ± 0.31	4.83 ± 1.74	1.86 ± 0.66	0.17 ± 0.09	9.05 ± 3.65
March 28, 2012	16	2.86 ± 0.48	4.68 ± 2.62	1.64 ± 0.71	0.20 ± 0.09	12.0 ± 5.41
April 24, 2012	12	3.26 ± 0.58	5.20 ± 2.47	1.81 ± 1.03	0.15 ± 0.14	8.73 ± 5.31
May 28, 2012	12	3.27 ± 0.56	5.20 ± 2.47	2.90 ± 3.50	0.14 ± 0.13	8.10 ± 5.25
June 28, 2012	17	3.17 ± 0.16	5.88 ± 0.91	2.19 ± 0.55	0.17 ± 0.11	7.25 ± 3.69
July 21, 2012	23	3.25 ± 0.43	5.98 ± 1.67	2.21 ± 0.78	0.16 ± 0.11	7.19 ± 4.51
August 03, 2012	12	3.95 ± 0.50	9.49 ± 3.91	3.47 ± 1.50	0.30 ± 0.28	7.56 ± 5.37
September 11, 2012	20	3.43 ± 0.44	5.43 ± 2.19	1.89 ± 0.91	0.28 ± 0.29	11.5 ± 7.43
October 05, 2012	21	3.71 ± 0.92	7.07 ± 5.38	2.86 ± 2.01	0.65 ± 0.73	18.7 ± 8.77
November 25, 2012	27	3.96 ± 0.49	8.75 ± 3.21	3.73 ± 1.47	1.31 ± 0.73	31.8 ± 7.72
December 30, 2012	19	3.96 ± 0.83	10.2 ± 8.64	4.10 ± 3.77	1.52 ± 1.94	28.4 ± 3.75

Table 1.
Average size and GSI of sampled opihi for the reproductive cycle study from November 2011 to December 2012; data values from n individuals are presented as mean ± SD.

2.2 Gonad development stage

The maturity stages of gonad development of Hawaiian limpet *C. sandwicensis* were reported and classified in **Table 2** and **Figure 2**. Multiple development stages (**Figure 2A**) were observed in the same ovary of female during the final maturation season. **Figure 2D** showed resting stages (April to August), because oocytes were not of clear formation from the ovary cell wall. Similar observation was made for gonad of the male (**Figure 2E**). The testes were densely packed with spermatozoa which appeared as dark blue stained by hematoxylin. Sperm were less densely in the male gonad (**Figure 2F**).

2.3 Sexual determination

Sexual determination of all limpet species is not known from external morphology. Render and maturation status of any limpet species could only be sexed after killing and dissecting. Our efforts were trying to examine ripeness of live animal

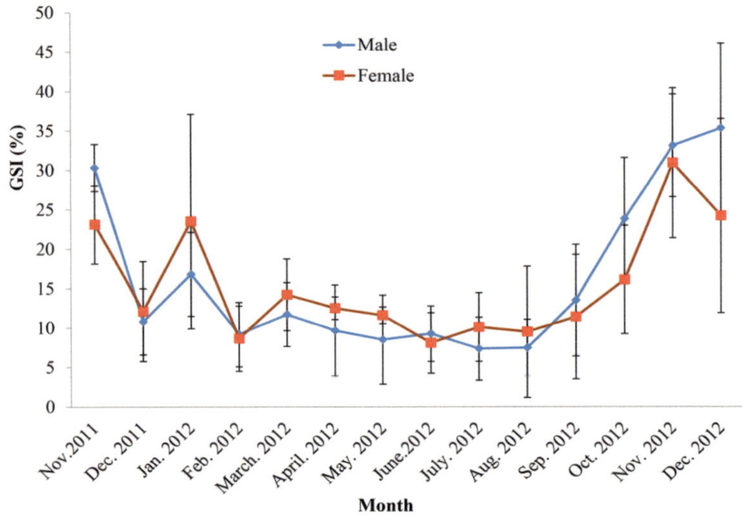

Figure 1.
Seasonal changes in GSI of males and females limpet C. sandwicensis.

Stage	Description	
1	Resting stage	The gonad is characterized by little or germinal epithelium, unclear distinguishable from ovary wall cells and also for spermatid, the initial oocytes about 2 μm
2	Early development	Nucleus enlarged, oocyte diameter about 7–10 μm. The male gonad is shaped like around tubule and a thick germinal epithelium lines the edges of the testes lobes
3	Late development	The ovaries are swollen laterally and some oocytes in the final stages of vitellogenesis. Cytoplasm granular, the oocytes diameter ranging from 50 to 100 μm
4	Ripe	Ovaries are swollen with dark brown color. Oocytes diameter ranging from 110 to 130 μm. The testis is dense with spermatozoa, milky white and/or dark blue stained by hematoxylin
5	Spawning and reabsorbing	Spawning testis contained about 80% mature sperm; the ovary contains less densities of mature oocytes relative to ripe gonads.

Table 2.
Maturation stages of gonad of limpets [10–12].

without killing them. We eventually found a way to assess the gonad, by placing them upside down on a table or putting them close to the edge of a substrate. When they try to attach to the substrate, they move their foot toward the substrate, and sometimes the gonads may be seen from the top of their head. Males were identified if the animals had milky white gonad near the edge of the shell, as shown by the arrow in **Figure 3A**. The gonad of the ripe female was dark brown or dark green in color (**Figure 3C**). It is noticed that this way, it can only be conducted when the animal reaches maturity stage or during the spawning season. Ultrasound was also an option method, but it's inconvenient and is not a practical way.

2.4 Fecundity and maturity size

Absolute fecundity (F) of mature female (n = 5) limpet *C. sandwicensis* is varied, and related maturity stage and body weight of animal found that total

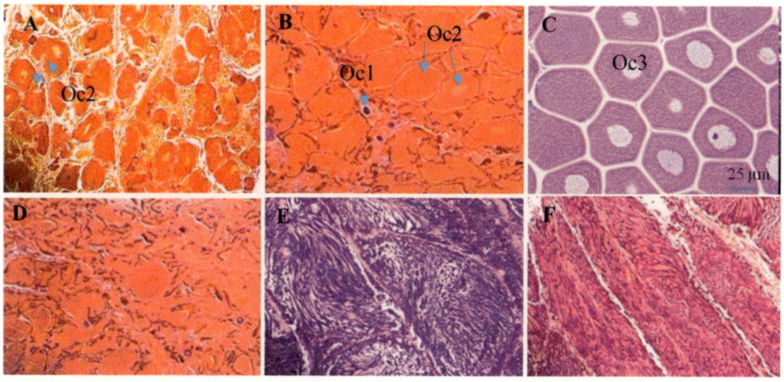

Figure 2.
Cross sections showing stages of limpet C. sandwicensis *gonad development. (A) Most oocytes in early development stages in ovary of female, (B) oocytes deforming shape in the ovary, (C) ripe stage and some oocytes were still in late development stage, (D) resting development stages of female, (E) mature male gonad with dark blue stained by hematoxylin, and (F) spermatic in gonad of male.*

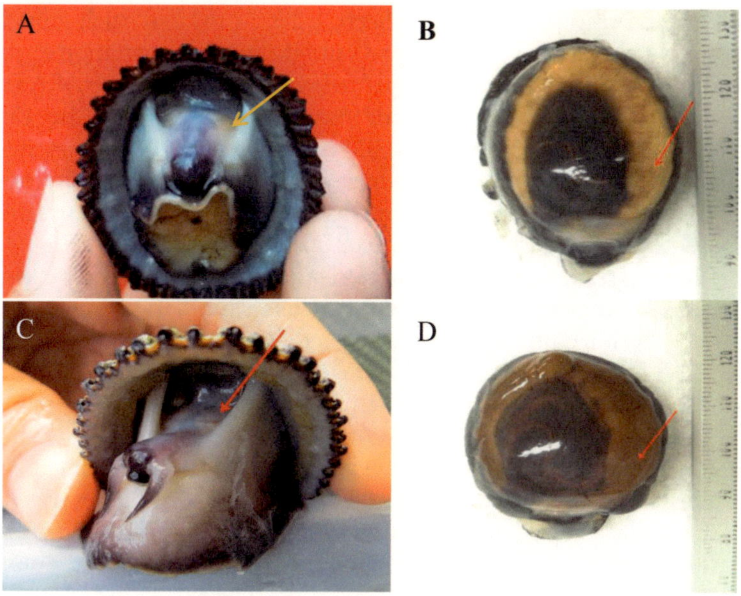

Figure 3.
Determination of sexually mature male and female limpet. (A) Live limpet before dissecting, the arrow shows a sign where mature male gonad could be identified; (B) soft body tissue was removed off the shell, the mature male gonad (milky white) took up all the space around the digestive gland (dark color). This supports the location of mature gonad where it was seen as pointed in picture A; (C) live female limpet before dissecting, the arrow points where mature female's gonad would be seen; (D) showing a dark brown mature female gonad with shell removal.

eggs counted ranged from 42.080 to 157.000 eggs per g body weight (BW). Fecundity was plotted against body weight with linearly correlated to body weight (P = 0.019), and the best is described by the equations F = 28.4 BW − 77.3 (R^2 = 0.96). The monthly recorded GSI data combined with histology analysis showed that the Hawaiian limpet *C. sandwicensis* would attain sexual maturity size about 1.5–2.0 cm in shell length. Other studies also found that *C. sandwicensis* attained reproductive stage at shell length of 2.3–2.5 cm or larger [2].

3. Seed production

3.1 Maturation culture

Two experiments were conducted to induce final maturation of limpet *C. sandwicensis* in the laboratory conditions. The first trial is formulated feed with the addition of arachidonic acid (ARA) into diet. The experimental diet is described and shown in the previous studies [12, 13]. In brief, limpet was fed with three diets including: control diet (without additional ARA), diet 2 containing 0.20% ARA, and diet 3 (0.33% ARA). Nice adult *C. sandwicensis* (3.07 ± 0.22 cm in shell length) species were fed with these diets for 90 days. Each limpet was randomly placed into its own colander of 20 cm diameter. The colanders were placed in aquaria (150 L) with a recycled water flow rate (15 L min^{-1}). Seawater was exchanged weekly of about 30%. The experiment was conducted under ambient photoperiod and temperature ranging from 23 to 25°C. Salinity was maintained at 35. Prior to the beginning of the experiment, several limpets were randomly selected among the group and dissected to obtain initial GSI and gonad development status. During the experimental period, three animals were randomly examined monthly to assess maturation status as described in Section 2.3. At the end, their gonads were extracted and weighed to obtain the gonad's weight for the calculation of the GSI.

The result showed that gonad of limpet fed with diet containing ARA increased three times higher than the GSI of animals that fed with the control diet (**Table 3**). There was a significant difference ($P < 0.05$) in GSI of animal that fed with diet incorporated with ARA as compared to those fed with control diet. There was no significant difference in GSI of those limpets fed with both diets 0.2% ARA and 0.33% ARA with the same ARA/EPA ratio of 0.70.

In the following trial, the final maturation of limpet was induced by using OvaRH (Syndel Laboratories Ltd. Canada) which is a synthetic salmon GnRH analog (sGnRHa). The hormone was injected directly into the gonad of limpets. Twelve limpets (9.17 ± 3.17 g/ind.) were tagged and weighed. Each limpet received a total of five to seven injections, at 7-day intervals at dose of 250 ng/g body weight (BW). The control treatment was run without hormone injection. During the period, experimental limpets were held on biofilm aquaria with water movement by an aquarium biofilter pump (567 L per hour). The maturation of limpet was examined weekly by randomly selected and sacrificed two limpets in each treatment. Their gonads were collected for calculation of GSI, and a piece of gonad was immediately fixed in 10% formalin for histological examination. The experiment was conducted during the final maturation and spawning season.

Day	Parameter	Control	0.2% ARA	0.33% ARA
Initial	GSI (%)	3.10 ± 2.48	3.10 ± 2.48	3.10 ± 2.48
45	GSI (%)	5.94 ± 5.65	11.0 ± 6.82	8.13 ± 0.52
	Egg size (μm)	—	118 ± 9.71^{a}	121 ± 9.42^{a}
75	GSI (%)	6.11	24.5	23.7
	Egg size (μm)	—	123 ± 4.23^{a}	121 ± 5.93^{a}
95	GSI (%)	4.21 ± 0.82^{a}	10.8 ± 4.47^{b}	15.5 ± 5.47^{b}

^{a,b} *The same letters in the row indicate no significant difference in eggs sizes, the empty grids indicate no egg was observed.*

Table 3.
Gonadal somatic index and egg size of limpet fed different dietary ARA for 95 days.

The results showed that the gonad of limpet was rapidly increased after 3 weeks with three injections (**Figure 4**). The GSI of limpet *C. sandwicensis* increased rapidly from initial 12.0–28.3% after the third injection and reached 32.9% for the final maturation stage after the fourth injection. GSI of limpets in the control group remained the same until finishing the experiment.

It is reported that the reproduction of aquatic animal is controlling by external and internal parameter factors such as photoperiods, food availability and hormone regulation. Therefore, our study focused on three factors including photoperiod, nutrient requirement as the ratio of highly unsaturated fatty acid arachidonic acid (ARA) and eicosapentaenoic acid (EPA), and gonadotropin-releasing hormone. In our hands photoperiod is important for the maturation of limpet. The effect of photoperiod may be seen more clearly in the following maturation trials when the experiment was run before and during the spawning season [11, 12]. This showed the role of environmental conditions in the regulation of the timing of the reproductive cycle of limpet. For example, the limpet *C. exarata* was found that the reproductive resting phase coincided with day length above 13 h [10], suggesting that a higher 13 h day length could inhibit gametogenesis of limpet. Photoperiod has also been reported to be influential on reproductive cycles of many marine invertebrates [14, 15].

Final maturation of *C. sandwicensis* was successfully induced by the addition of arachidonic acid (ARA) into diet to obtain an appropriate ARA per eicosapentaenoic acid (EPA) ratio. Arachidonic acid serves as a precursor for the synthetic of prostaglandins which are functional for reproductive process [16]. Prostaglandins play a critical role during the ovulatory process in teleost fishes [17]. Our previous study found that *C. sandwicensis* preferred to feed on benthic diatoms in the wild [18]. Several benthic diatoms such as *Nitzschia*, *Amphora*, and *Navicula* were predominant in the stomach content of *C. sandwicensis*, and literature studies found that these diatoms contained high level of ARA and EPA [19–21]. Our review found that an ARA/EPA ratio of about 0.70 was found in several benthic diatoms such as *Nitzschia* and *Chaetoceros* suggesting that this would be a good starting point for experimental diet on adult limpet *C.*

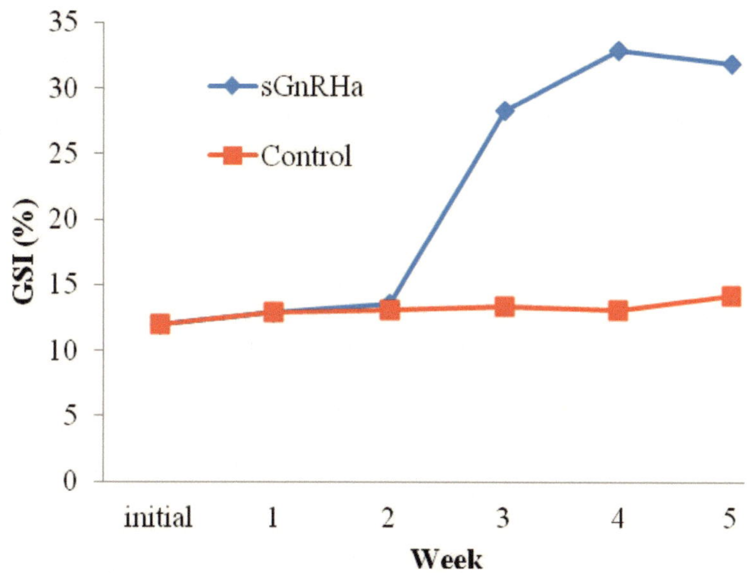

Figure 4.
Gonadosomatic growth of limpet C. sandwicensis *by hormone induction.*

sandwicensis. Thus, experimental trial on different ARA to EPA ratios of 0.70 was conducted; as a result, *C. sandwicensis* reached final maturation [12]. This result provided significant data on the effect of ARA/EPA on maturation of limpet and gastropods as well.

GnRH-like peptides that existed in the central nervous system and peripheral chemosensory organ of sea hare *Aplysia* were detectable by antisera against mGnRH [22]. These GnRH-like peptides controlled egg laying of *Aplysia*. For abalone, studies had demonstrated the existence of GnRH-like peptides in the neural ganglia and ovary of abalone [23, 24], and the existence of GnRH-like peptide in the neural ganglia was determined by using immunohistochemistry and reverse-phase high-performance liquid chromatography [24, 25].

The Hawaiian limpet *C. sandwicensis* were also induced to final maturation using salmon GnRH analog (sGnRHa) at dose of 250 ng/g BW. The sGnRHa stimulated gonad development and final maturation in limpet in 5 weeks when they injected at 7-day intervals at low concentration 250 ng/g BW [13]. The GSI increased significantly from the third week of injection and developed rapidly and reached to the maximum level after 4 weeks of injection as compared to the control, which did not show gonadal development (**Figure 4**). This shows that GnRH also involved in regulating reproductive development in limpet. Similar finding was also reported in abalone; the adult abalone was induced to final maturation in 5 weeks by weekly injection of these GnRHs at low dose (250 ng/g BW) and induced spawning at higher dose of 1000 ng/g BW [23]. The existence of GnRH-like peptides in the neural ganglia and ovary of the abalone [23, 24] and the existence of GnRH-like peptide in the neural ganglia were determined by using immunohistochemistry and reverse-phase high-performance liquid chromatography [24, 25]. GnRH-like peptides that existed in the central nervous system and peripheral chemosensory organ of sea hare *Aplysia* were detectable by antisera against mGnRH [22]. These GnRH-like peptides controlled egg laying of *Aplysia*. The mammalian GnRH analog was known to stimulate maturation and induced spawning in abalone [23]. The responses of molluskan to environmental cues are controlled by hormones, and the principal sources of hormones within molluscan nervous system are neurosecretory cells [26]. Our results suggest that diatom blooms may be the environment cues. GnRH could stimulate reproductive process by acting directly on the gonad in limpet. Both limpet and abalone are marine gastropod species. This process would be also facilitated by the reproductive photoperiod, and/or the right photoperiod would stimulate the increased secretion of luteinizing hormone and follicle-stimulating hormone that enhances the reproductive process in limpet *C. sandwicensis*.

3.2 Spawning induction

Two different spawning methods were conducted to examine the optimal method of spawning for the Hawaiian limpet. The first method was conducted using hydrogen peroxide. Hydrogen peroxide is a traditional method used for spawning induction in abalone. **Figure 5** shows the addition of H_2O_2 to seawater is believed to produce hydroperoxy free radicals (HOO^-) and peroxy radicals (OO^-); these radicals of activated oxygen suitable for the cyclooxygenase catalyzed addition of prostaglandin [27–29].

Experimental animal. Limpet broodstock (>3.0 cm in shell length) were collected at the shoreline from a remote area on Oahu island. They were held on biofilm aquaria for 2 days before use for the experiment. Sexually matured broodstock were selected as described in Section 2.3. Eight matured limpets

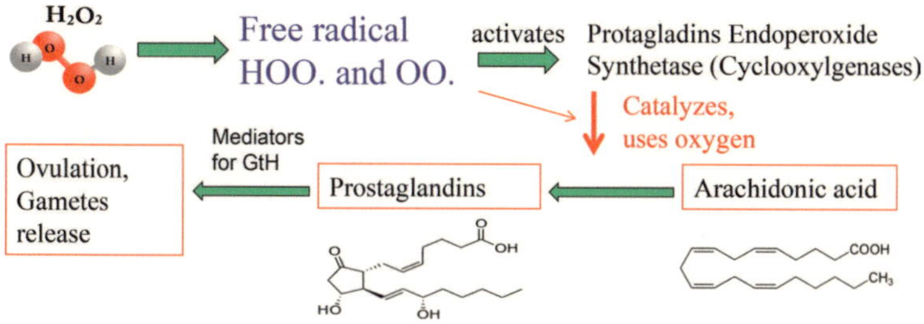

Figure 5.
Mechanism of hydrogen peroxide in spawning induction of mollusk species.

(approximate sex ratio, male:female = 1:1) were selected from the holding tank then placed into a spawning container (3 L) with fresh clean seawater for each trial. The spawning container was gently aerated, and pH in the spawning container was first adjusted to pH 9–9.5 by 1 M of Tris-base for about an hour. Thereafter, stock 6% of H_2O_2 was slowly added to spawning container to obtain the desired concentration. The broodstock were exposed to five different concentrations of H_2O_2. These are control (without H_2O_2), 0.6×10^{-2}%, 1.20%, 1.49×10^{-2}%, and 1.80×10^{-2}%. The exposing time ranged from 5 to 45 min depending on the response of animals. Then the spawning activities were observed. The results show that the highest number of spawners was induced at 0.6×10^{-2}% and no mortality occurred in the 24 h after spawning. Most of animals died at 1.49×10^{-2}% and 1.80×10^{-2}% in the 24 h after exposure to these levels. These results highlighted the nonspecific toxic effect of the chemical. Similarly, at this level all animals were dead eventually, but this level induced 10–15% spawning [10]. However, we concluded that this method may not be used as a practical method and not recommended for spawning induction in limpet. This was due to a nonspecific effect, and the broodstock eventually died within a week after being exposed to H_2O_2. The second method with GnRH at dose of 1000 ng/g BW may be considered as the most practical induction spawning method for limpet because there were no mortalities occurring after spawning.

This could probably be due to the instability of H_2O_2. The H_2O_2 was fresher, and we ordered before use. No mortality occurred in the 24 h after spawning at 0.6×10^{-2}%. This led to the thought that 0.6×10^{-2}% may be safe, but in the last trial at this concentration, all animals died within a week after exposure to H_2O_2. We used this level in further spawning the trials. The limpet may have released gametes because they thought they were dying. This is a well-known phenomenon among fruit trees that are sometimes even sprayed with herbicide to get them to fruit. Under the microscope we found that a high percentage of immature eggs with different sizes, these eggs were not successfully fertilized. This concluded that hydrogen peroxide is not a practical method.

Induction of spawning by using sGnRHa is an applicable technique and was the most practical method. There were no mortalities after injection of sGnRHa, and 100% animals survived after spawning. However, it is noticed that spawning induction of limpet by GnRH is effective only on ripe *C. sandwicensis*.

3.3 Embryonic and larval development

Different larval development of *C. sandwicensis* is shown in **Table 4**. There were 18 distinct stages of larval development of *C. sandwicensis* in this study. Spawned eggs were 111 ± 5.64 μm (**Figure 6a**). The first polar body appeared in about 30–45 min after spawning indicating fertilized eggs (**Figure 6b**). The two-cell

Sequent stage	Embryo, larval development stage	Time (h)
1	Fertilization	0.00
2	Discharge of first polar	0.30–0.45
3	First cleavage (2 cells)	1.00–1.30
4	Second cleavage (4 cells)	2.00–2.30
5	Third cleavage (8 cells)	3.00–3.30
6	Morula	3.30–4.00
7	Blastula	4.00–4.30
8	Gastrula	4.30–5.00
9	Appearance of cilia forming prototrochal	8.00–10.00
10	Trochophore ready to hatch out	10.30–11.30
11	Trochophore free swimming larvae	12.00–14.00
12	Continue extended cilia	13.30–14.30
13	Completion of girdle and cilia develop	14.30–16.00
14	Larval shell formation	14.30–16.00
15	Advance larvae shell formation	16.30–18.00
16	Exhibiting flat apical from larval shell and complete developed velum and cilia	18.00–20.00
17	Eye spot	20.00–21.30
18	Completed muscle formation	21.30–24.0

Table 4.
Embryonic and larval development of limpet C. sandwicensis *at (22°C).*

stage (stage 3) was found within 2 h after spawning (**Figure 6b**). About 10 h post-fertilization, protrochophore stage with cilia appeared (**Figure 6b**). Larvae started hatching out at 12–14 h. The length and width of free swimming larvae were 85.5 ± 9.5 μm and 79.6 ± 7.9 μm, respectively. Larvae continued to develop velum from cilia, and apical region became flat for shell formation in about 18–20 h after spawning (**Figure 6o** and **p**).

3.4 Larval rearing

Several studies [10, 11] on settlement of *C. sandwicensis* larvae on different combinations of diatom and pelagic algae were conducted. The results showed that mixture of diatom *Amphora* and pelagic *Palova* induced the highest survival rate (21.7 ± 7.07%) of settled larvae. Diatom *Nitzschia* seemed not to be preferred by *C. sandwicensis* larvae because the observation noticed that high mortalities occurred from 4 to 6 days. Pelagic algal *Palova* may be preferred over *Isochrysis*. Among the surviving larvae, all of them settled after 3 days and fed on diatoms. On the other hand, different plate substrates reported to be affected on larval settlement of gastropod species, such as abalone of roughened plexiglass, and corrugated plastic sheet, and the rubberized canvas seemed to be preferred for settlement over fibrocement board. The results of our study were higher than previous study which was attempting to induce the settlement of Hawaiian limpet *C. sandwicensis* larvae on different substrata [10]. She found that mylar plastic and plexiglass induced a significantly higher larval settlement compared to glass, smooth and rough basalt

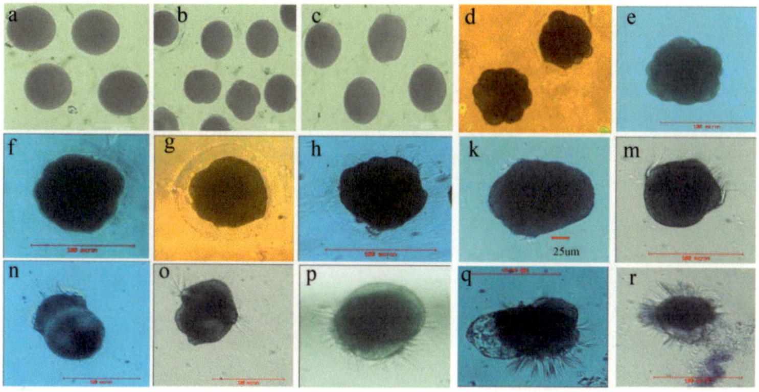

Figure 6.
Embryonic development stage of Hawaiian limpet C. sandwicensis. *(a) Spawned egg and stage 1 spermatozoids; (b) stages 2 and 3, discharge of polar body and first cleavage (2 cells), and stage 4, second cleavage (4 cells); (c) stage 5, third cleavage (8 cells); (d) stages 6–7 morula and blastula; (e) stage 8, gastrula; (f and g) stage 9, appearance of cilia forming the prototrochal; (h) stage 10, trochophore larvae ready for hatch out; (k) stage 11, trochophore larvae free swimming; (m) stage 12, trochophore free swimming larvae with extend cilia; (n) stage 13, complete girdle, cilia develop and apical; (o) stage 14, early larvae shell formation; (p) stages 15 and 16, veliger larvae exhibiting flat apical from larval shell and complete developed velum and cilia; (q) stage 17, appearance of eye spot; (r) stage 18, appearance of muscle attached.*

rocks, coral skeleton, and textured and untextured plastic. However, the settlement rates were very low ranging from 1.58 to 7.73%. It could probably be due to inappropriate benthic diatoms.

The role of benthic diatoms *Navicula*, *Amphora*, *Nitzschia*, and others were reported to best diatom species induced the settlement and metamorphosis of abalone larvae [19, 30, 31]. The effects of different benthic diatoms grown on different plate substrates on metamorphosis of the tropical abalone *Haliotis asinina* were reported by [31]. They found that mixture of diatoms induced significantly higher metamorphosis rate of abalone larvae than other group including *Amphora*, *Amphora* + *Nitzschia*, and *Nitzschia* with any plate substrate. This suggested that mixture of benthic diatoms is better than single once. Another study also found that a mixture of benthic diatoms consisting of *Navicula* and *Amphora* produced a significantly higher growth and survival rates for abalone larvae *H. discus hannai* than monocultures benthic diatoms [19]. The report showed that the monocultures of benthic diatoms produced a poor growth and did not support survival for more than 2 weeks especially *Nitzschia*. The authors also stated this could be due to the difference in nutritional value of these benthic diatoms. In particular, the EPA value in *Navicula* and *Amphora* was reported to be higher than the value in *Nitzschia* [19]. These results support our study that mixture of diatom and pelagic algae induced better survival rate of Hawaiian limpet and mixture of diatom *Amphora* and pelagic *Palova* would be recommended for future use of larval rearing of the *C. sandwicensis*.

4. Culture system

There is a lack of study on aquaculture system for limpet species as well as the Hawaiian limpet *C. sandwicensis*. We have attempted to raise the limpet *C. sandwicensis* in system with water flow through, but transfer mortality is a challenge because the animals cling tightly to the cultured tank walls. It was hard to get them off the wall without injury. Later we found that putting plastic sheets as tank liners solves this problem (**Figure** 7).

Figure 7.
A circular holding biofilm tank without plastic liner, and three aquaria with plastic sheer liner above, used for the second and following holdings.

Rocky habitat and adhering to the substrates are problems. Limpet *C. sandwicensis* attach to the washing rocks in the wild. They cling to the culture tank with their muscular foot. It indicates that physical damaged may happen while removing them off the tank's wall. Similar observation has been made in abalone; they often succumb to wound suffered during removal off the substrates. Abalone blood has no clotting ability, and relatively minor cut can cause death due to loss of hemolymph [32]. Eventually, we developed plastic tank liners that were our breakthrough for transferring animals from one tank to another. Our study was the first to reveal that the Hawaiian limpet *C. sandwicensis* was healthy and fed well in the experimental aquaria without intermittent water sprayed or dump tanks.

5. Feed development and nutrient requirement

5.1 Development of formulated feed

We [18] began our studies in this area with several preliminary tests on biofilm because *C. sandwicensis* ate biofilm well which should be close to their natural diet. We also tested several dry diets, gelatin, and agar diets. We discovered that several were preferred and some were not. Several chemical attractants were tested including betaine, gamma aminobutyric acid (GABA), and dimethyl propiothetin (DMPT), but these did not enhance feeding. Among the feeds tested in a preliminary way, fish meal and soybean meal as well as feeds incorporating biofilm were preferred. Eventually, we found that *Porphyra* preparations could replace biofilm as a feeding stimulant in formulated feed.

Nutrient requirement was our next step to develop the commercial feed available for limpet, and the authors assumed that the nutrient requirement of limpet and abalone is the same as they are marine gastropod [18]. For abalone, a series number of researches had been done, and the optimum nutrient requirement as protein, carbohydrate, and lipid was focused. However, the results still varied among researchers. For example, the protein requirements of abalone found by previous studies [33, 34] were higher than those reported in the previous studies [35–37]. Poor growth was found for abalone when the animal was fed with formulated diet containing amino acid profile that does not match the animals' tissue [38]. Moreover, other studies [39, 40] found that a significant lower growth rates when abalone fed with dried kelp *Ecklonia maxima* and *Laminaria*. Therefore, these studies raised the hypothesis that the growth rate of abalone is related to the degree of the amino acid profile of feed and the amino acid profile of tissue.

5.2 Protein and carbohydrate requirement

Based on the results of previous study [18], further studied on the determination of protein and carbohydrate requirement for Hawaiian limpet *C. sandwicensis* [41].

Experimental animals. Adult *C. sandwicensis* limpets (shell length above 3.0 cm) were collected from a remote area in Oahu, Hawaii, used for this study. After collection they were immediately placed into a 14 L ice plastic insulation box with plastic liner and then transported to the laboratory at the University of Hawaii in Monoa. The limpet was held in a plastic aquaria 150 L with water flow for a week; during this period, the animal were fed with the experimental diet and the commercial algae *Porphyra tenera* or *yezoensis*, known as Nori (Nishimoto Trading Co. Ltd., Korea).

Experimental diets. Formulations of dietary protein and carbohydrate levels are shown in **Table 5**. The first trial was done for dietary protein level, following by dietary carbohydrate. For carbohydrate trial, four different dietary carbohydrate levels of 18, 27, 32, and 37% were tested. The amino acid profiles of *C. sandwicensis* tissue and of the dietary protein in trial 1 were analyzed at the Aquatic Feed and Nutrition Laboratory, Oceanic Institute, Hawaii, USA, according to the described method [42]. The results are presented as A/E ratio (**Table 6**). Most of the essential amino acids of diets were identical and/or close to the amino acid profile of *C.*

Ingredient	Dietary protein trial 1					Dietary protein trial 2				Carbohydrate trial			
	270	320	370	420	470	210	300	350	500	180	270	320	370
Fishmeal	16.5	19.5	22.5	25.5	28.5	13.4	17.0	21.0	30.4	16.5	16.5	16.5	16.5
Defatted soybean	11.5	14.5	17.5	20.5	23.5	11.0	12.7	16.6	24.4	11.5	11.5	11.5	11.5
Krill meal	4.5	7.5	10.5	13.5	16.5	7.1	8.0	11.0	16.1	4.5	4.5	4.5	4.5
Porphyra[1]	14.0	14.0	14.0	14.0	14.0	14.0	14.0	14.0	14.0	14.0	14.0	14.0	14.0
Wheat flour	15.4	14.3	13.3	12.2	11.1	8.98	5.3	4.3	0.8	15.4	26.9	33.7	40.5
Diatomaceous earth	29.2	21.9	14.6	73.0	0.0	36.8	35.2	25.8	7.3	30.9	19.4	12.6	5.8
Alginate	5.0	5.0	5.0	5.0	5.0	5.0	5.0	5.0	5.0	5.0	5.0	5.0	5.0
Corn/fish oil[2]	2.5	1.9	1.2	0.6	0.0	2.32	1.4	0.9	0.6	1.0	1.0	1.0	1.0
Vit. mix[3]	1.0	1.0	1.0	1.0	1.0	1.0	1.0	1.0	1.0	1.0	1.0	1.0	1.0
Cholesterol	0.4	0.4	0.4	0.4	0.4	0.4	0.4	0.4	0.4	0.2	0.2	0.2	0.2
Total	100	100	100	100	100	100	100	100	100	100	100	100	100
Water	100	100	100	100	100	133	133	133	133	100	100	100	100
Analyzed and calculated nutrient "as fed"													
Crude protein	26.5	31.7	37.0	42.4	47.7	21.2	30.5	35.8	49.2	26.5	26.5	26.5	26.5
Crude lipid	4.97	4.97	4.97	4.97	4.97	5.13	5.13	5.13	5.13	3.47	3.47	3.47	3.4.7
Carbohydrate	17.5	17.8	18.1	18.3	18.0	11.0	11.0	11.2	11.4	18.0	27.0	32.0	37.0

[1]*This is commercial seasoned seaweed known as nori or the red algae Porphyra tenera or yezoensis. Nishimoto Trading Co. Ltd., Korea.*
[2]*Corn oil and menhaden oil (1:1; v/v).*
[3]*Commercial vitamin mix (NRC 1981) was kindly provided from Dr. Warren Dominy (Oceanic Institute).*

Table 5.
Composition of formulated diet (% dry matter).

Essential AA	Preliminary protein trial (trial 1)						Second protein trial			
	Tissue	270	320	370	420	470	210	300	350	500
Arg	224	123	118	129	125	139	208	198	190	173
His	33.8	39.5	37.4	41.5	39.9	44.6	29.1	29.4	29.8	32.3
Ile	81.4	88.7	89.5	87.7	88.6	99.5	80.4	80.9	81.2	80.8
Leu	146	158	157	155	156	177	149	148	147	142
Lys	69.2	159	170	149	160	178	110	112	114	116
Met/Cys	68.3	74.1	75.1	77.6	77.7	87.0	51.2	53.5	55.4	57.4
Phe/Tyr	123	158	161	159	160	179	187	194	200	219
Thr	136	97.9	92.0	97.8	93.4	98.0	78.9	79.1	79.3	78.4
Val	117	103	99.9	103	99.8	111	107	105	104	100

Table 6.
The A/E ratio [(each EAA/total EAA) × 1000] amino acids of dietary protein and animal tissue.

sandwicensis tissue except for Arg and Thr which were lower in the experimental diets compared to the tissue.

The process of feed preparation for extrusion of all diets was based on the methods described by the previous study [18, 41]. In brief, fish meal, soybean meal, and krill meal were mixed thoroughly with other ingredients. Wheat flour was used as starch, and diatomaceous earth was used as filter to balance in the diets. Wheat flour and alginate were gelatinized in boiling water (about 25% of total dried weight basis) before being mixed with other ingredients. Other ingredients were then mixed thoroughly with the gelatinized solution; thereafter the mixed (paste) was heated in boiled water bath again for about 2 min. The paste was shaped into sheets about 1.0 mm thickness and then cut into 1.2 cm^2/pieces and dried naturally in laboratory conditions for about 1–2 h. The pieces were then sealed in a plastic sample bag and stored at −4°C until use.

Each limpet was randomly placed into its own colander of 20 cm diameter (**Figure 8**). The colanders were placed in aquaria (150 L) with a recycled water flow rate (15 L min^{-1}). Nice limpets were used for diet, and the experiment was run for 90 days.

The growth of animals in weight (g) and shell length (cm) was measured monthly. The growth was expressed in terms of specific growth rate (SGR), weight gain, and shell length increasing. The shell length was measured with an electronic digital caliper (0.01 μm), and the weight was determined with an electronic scale (0.01 g error) for every 4 weeks:

$SGR = \{(\ln Wf - \ln Wi)/T\} \times 100$, where Wf is the final weight, Wi is the initial weight, and T is the total day of the experiment.

The result showed that the growth response of *C. sandwicensis* in terms of weight gain (%) of animals in dietary protein trial 2 was fitted into quadratic models (**Figure 9**). The best fit for the estimation of optimal protein level could be described as $Y = -0.0003x^2 + 0.234x - 21.8$ ($R^2 = 0.96$). The trend of growth showed that maximum weight gain appeared to be about 35% dietary protein.

The response of *C. sandwicensis* in weight gain to dietary carbohydrate levels was then fitted to quadratic models (**Figure 10**). It shows that the weight gains of *C. sandwicensis* progressively increased and reached their maximum value at a carbohydrate level of about 27%, which could probably be described as $Y = -0.0012x^2 + 0.64x - 56.7$ with the correlation value of $R^2 = 0.91$.

Figure 8.
Experimental colander with an C. sandwicensis *on it; a small square is a piece of feed.*

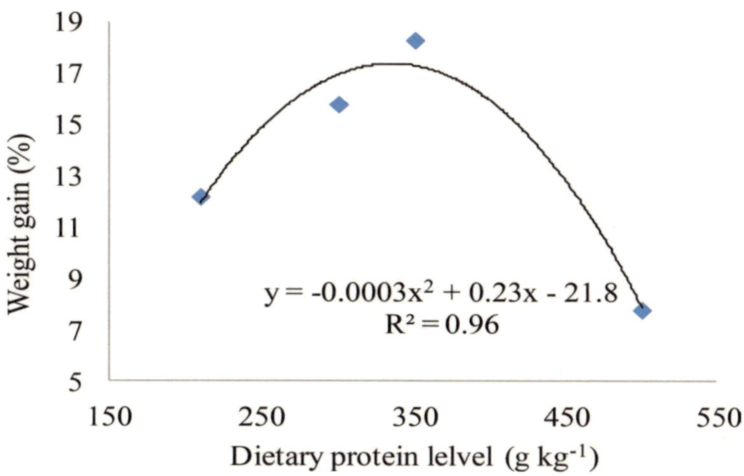

Figure 9.
Relationship between weight gain and dietary protein level of trial 2 for C. sandwicensis *for 60 days.*

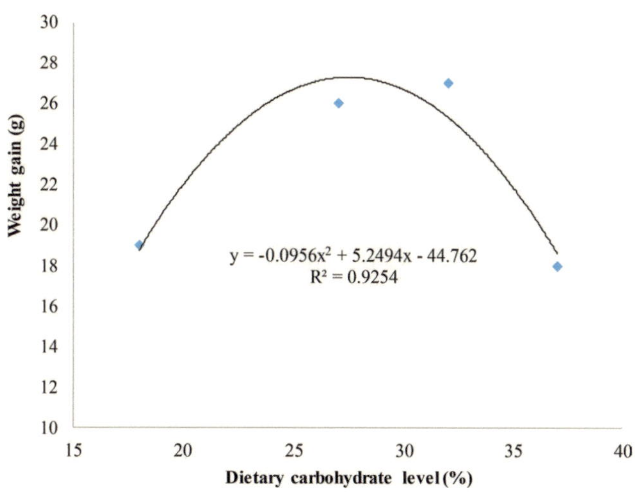

Figure 10.
Relationship between weight gain and dietary carbohydrate level for C. sandwicensis.

5.3 Energy requirement

Recent study found that limpet *C. sandwicensis* required no specific effect on dietary protein to energy (PE) ratio when the animal was offered with diet containing various PE ratios ranging from 87.2 to 102.9 mg/kcal [43]. There was no significant effect on growth performance of limpet among the diets, but a PE ratio of 87.2 mg/kcal produced the best tissue growth.

6. Conclusion

This chapter provides scientific basis for the development of aquaculture techniques for the Hawaiian limpet *C. sandwicensis*. Several reproductive characteristics of the Hawaiian limpet *C. sandwicensis* were investigated such as natural spawning season (November to January), maturity size (above 1.5 cm shell length), and gonad development stage (5 stages), examining sexually matured of male and female animals. The second important issue is seed production. Induction of final maturation using dietary ARA/EPA ratio of 0.70 and injection of sGnRH at dose of 250 ng/g BW is recommended. Induction of spawning by using sGnRHa is an applicable technique and was the most practical method with no mortality occurring. Pelagic algal *Palova* and benthic diatom *Amphora* induced good survival rate for larval settlement and are potential algal species for commercial hatchery of larval rearing for *C. sandwicensis*. An effective method of using plastic liner and/or colander for handling and potential use for culture system of limpet was also developed. A practical commercial feed with good palatability, producing good growth performance at 35% dietary protein, 32% for carbohydrate and protein to energy (PE) ratio ranging from 87.2 to 102.9 mg/kcal could be used for commercial limpet feed.

Acknowledgements

This work was financially supported by the Center for Tropical and Subtropical Aquaculture, USA (Grant #@2009-210), and the Vietnam Education Foundation.

Author details

Hua Thai Nhan[1*] and Harry Ako[2]

1 College of Aquaculture and Fisheries, Can Tho University, Can Tho City, Vietnam

2 Department of Molecular Biosciences and Bioengineering, University of Hawai'i at Manoa, USA

*Address all correspondence to: htnhan@ctu.edu.vn

References

[1] WoRMS, Adams CH. In: Bouchet P, Gofas S, Rosenberg G, editors. World Marine Mollusca Database; 2010. Accessed through: World Register of Marine Species at: http://www.marinespecies.org/aphia.php?p=taxdetails&id=205749 [Accessed: 31 October, 2010]

[2] Kay EA, Magruder W. The Biology of Opihi. Honolulu: Department of Planning and Economic Development; 1977

[3] Barber AH, Lu D, Pugno NM. Extreme strength observed in limpet teeth. Journal of the Royal Society Interface. 2015;**12**:20141326

[4] Corpuz GC. Laboratory culture of *Cellana exarata* reeve (Gastropoda: Prosobranchia, Patellidae). Aquaculture. 1981;**24**:219-231

[5] Bird CE. Aspects of community ecology on wave-exposed rocky Hawaiian coasts [Doctoral thesis]. Hawaii, USA: University of Hawai'i at Mānoa; 2006

[6] Bird CE. Morphological and behavioral evidence for adaptive diversification of sympatric Hawaiian limpets (*Cellana* spp.). Integrative and Comparative Biology. 2011;**51**(3):466-473

[7] Iacchei M. Managing fisheries for future. Ka Pili Kai. 2011;**33**:3-5

[8] Thompson D. The opihi shellfish story. The Honolulu Magazine. 2011. Available from: http://www.honolulumagazine.com/Honolulu-Magazine/November-2011/The-Opihi-Shellfish-Story/

[9] Handy ES, Handy E, Pukui MK. Native Planers in Old Hawai'i, their Life, Lore and Environment. Honolulu, Hawai'i: Bishop Museum Press; 1991

[10] Corpuz GC. Life history traits influencing vertical zonation in the Hawaiian intertidal species of *Cellana* [doctoral thesis]. Manoa, Hawaii, USA: University of Hawaii; 1983

[11] Nhan HT. Development of aquaculture technology for the Hawaiian opihi *Cellana* spp [doctoral dissertation]. Manoa, Hawaii, USA: Honolulu University of Hawaii; 2014

[12] Nhan HT, Ako H. Reproductive biology and effect of arachidonic acid level in broodstock diet on final maturation of the Hawaiian limpet *Cellana sandwicensis*. Journal of Aquaculture Research and Development. 2014;**5**:256-264

[13] Nhan HT, Ako H. Maturation and spawning induction in Hawaiian opihi *Cellana* spp. by hormone GnRH. Communications in Agricultural and Applied Biological Sciences. 2013;**78**:194-197

[14] Giese AC, Pearse JS. Introduction: General principles. In: Giese AC, Pearse JS, editors. Reproduction of Marine Invertebrates. Vol. 1. New York: Academic Press; 1974. pp. 1-49

[15] McClintock J, Watts S. The effects of photoperiod on gametogenesis in the tropical sea urchin *Eucidaris tribuloides*. Journal of Experimental Marine Biology and Ecology. 1990;**139**:175-184

[16] Schimz G, Ecker J. The opposing effect of n-3 and n-6 fatty acids. Progress in Lipid Research. 2008;**47**:147-155

[17] Goezt FW. Hormonal control oocyte final maturation and ovulation of fishes. In: Ho WS, Randall DJ, editors. Fish Physiology. Vol. 9B. New York, NY: Academic Press; 1983. pp. 117-169

[18] Nhan HT, Ako H. Enabling studies for aquaculture of the Hawaiian Opihi, the limpet *Cellana*. World Aquaculture Society. 2012;**12**:40-44

[19] Gordon N, Neori A, Shpigel M, Lee MJ, Harpaz S. Effect of diatom diets on growth and survival of the abalone *Haliotis discus hannai* postlarvae. Aquaculture. 2006;**252**:225-233

[20] Ako H. Algae in the aquaculture of marine fishes and shrimps. In: Sustainable Aquaculture 95, Proceedings of the Pacific Congress on Marine Science and Technology, PACON; 1995. pp. 7-14

[21] Volkman JK, Jeffrey SW, Nichols PD, Rogers GI, Garlandv CD. Fatty acid and lipid composition of 10 species of microalgae used in mariculture. Journal of Experimental Marine Biology and Ecology. 1989;**128**:219-240

[22] Tsai PS, Maldonado TA, Lunden JB. Localization of gonadotropin-releasing hormone in the central nervous system and a peripheral chemosensory organ of *Aplysia californica*. General and Comparative Endocrinology. 2003;**130**:20-28

[23] Nuurai P, Engsusophon A, Poomtong T, Sretarugsa P, Hanna P, Sobhon P, et al. Stimulatory effects of egg-laying hormone and gonadotropin-releasing hormone on reproduction of the *Tropical Abalone*, *Haliotis asinina* Linnaeus. Journal of Shellfish Research. 2010;**29**(3):627-635

[24] Amano M, Moriyama S, Okubo K, Amiya N, Takahashi A, Biochemical OY. Immunohistochemical analyses of a GnRH-like peptide in the neural ganglia of the Pacific abalone *Haliotis Discus Hannai* (Gastropoda). Zoological Science. 2010;7(8):656-661

[25] Hahn KO. The neurosecretory staining in the cerebral ganglia of the Japanese abalone (Ezo-awabi), *Haliotis discus hannai*, and its relationship to reproduction. General and Comparative Endocrinology. 1994;**93**:295-303

[26] Joosse J. Evolutionary aspects of endocrine system and the hormonal control of reproduction of mollusk. In: Barrlington EJW, editor. Hormones and Evolution. Vol. I. New York: Academic Press; 1979. pp. 119-157

[27] Morse DE, Duncan H, Hooker N, Morse A. Hydrogen peroxide induces spawning in mollusks, with activation of prostaglandin endoperoxide synthetase. Science. 1977;**196**:298-300

[28] Morse DE. Biochemical and genetic engineering for improved production of abalone and other valuable mollusks. Aquaculture. 1984;**39**:263-282

[29] Moss GA, Illingworth J, Tong LJ. Comparing two simple methods to induce spawning in the New Zealand abalone (Pauna) *Haliotis iris*. New Zealand Journal of Marine and Fresh Water Research. 1995;**29**:329-333

[30] Searcy-Bernal R, Salas AE, Flores-Aguilar RA, Hinojisa-Rivera PR. Simultaneous comparison of methods for settlement and metamorphosis induction in the red abalone *Haliotis rufescens*. Aquaculture. 1992;**105**:241-250

[31] Gapasin RSJ, Polohan BB. Response of the tropical abalone, *Haliotis asinina*, larvae on combinations of attachment cues. Hydrobiologia. 2005;**548**:301-306

[32] Cox KW. Californian abalones, family haliotidae. California Fish and Game Fisheries Bulletin. 2005:118-133

[33] Britz PJ. Effect of dietary protein level on growth performance of South African abalone *Haliotis midae*, fed fishmeal-based semi-purified diets. Aquaculture. 1996;**140**:55-61

[34] Gomez-Montes L, Garcia-Esquivel Z, D'Abramom LR, Shimada A, Vasquez-Pelaez C, Viana MT. Effect of dietary protein: Ratio on intake, growth and metabolism of juvenile green abalone *Haliotis fulgens*. Aquaculture. 2003;**220**:769-780

[35] Mai K, Mercer JP, Donlon J. Comparative studies on the nutrition of two species of abalone, *Haliotis tuberculata* L. and *Haliotis discus hannai* Ino. IV. Optimum dietary protein level for growth. Aquaculture. 1995;**136**:165-180

[36] Coote TA, Hone PW, Van Barneveld RW, Maguire GT. Optimal protein level in a semipurified diet for juvenile greenlip *Haliotis laevigata*. Aquaculture Nutrition. 2000;**6**:213-220

[37] Sales J, Truter PJ, Birtz PJ. Optimum dietary crude protein level for growth in South African abalone (*Haliotis midae* L.). Aquaculture Nutrition. 2003;**9**:85-89

[38] Bautista-Teruel M, Fermin AC, Koshio SS. Diet development and evaluation for juvenile abalone, *Haliotis asinina*: Animal and plant protein sources. Aquaculture. 2003;**219**:645-653

[39] Naidoo K, Gavin M, Ruck K, Bolton JJ. A comparison of various seaweed-based diets and formulated feed on growth rate of abalone in a land-based aquaculture system. Journal of Applied Phycology. 2006;**18**:437-443

[40] Cho SH, Park J, Yoo JH. Effect of casein substitution with fishmeal, soybean meal and crustacean meal in the diet of the abalone *Haliotis discus hannai* Ino. Aquaculture Nutrition. 2008;**14**:61-66

[41] Hua NT, Ako H. Dietary protein and carbohydrate requirement of juvenile Hawaiian limpet (*Cellana sandwicensis* Pease, 1861) fed practical diet. International Aquatic Research. 2016;**8**(4):323-332

[42] Ju ZY, Forster I, Conquest L, Dominy W, Kuo WC, Horgen FD. Determination of microbial community structures of shrimp floc cultures by biomarkers and analysis of floc amino acid profiles. Aquaculture Research. 2008:1-16

[43] Mau A, Jha R. Effects of dietary protein to energy ratios on growth performance of yellowfoot limpet (*Cellana sandwicensis* Pease, 1861). Aquaculture Report. 2018;**10**:17-22

Chapter 6

Sponge Fishery and Aquaculture in Cuba: Impacts and Challenges

Betanzos-Vega Abel, Mazón-Suástegui José Manuel, Formoso-García Mario and Avilés-Quevedo María Araceli

Abstract

Sponges are very primitive multicellular organisms that belong to *phylum Porifera*; they are sessile and live attached to different types of hard and soft substrates. Sponges have different shapes and colours and very varied sizes, from a few millimetres to more than 2 m in height. They inhabit mainly in the marine environment at different depths. This chapter describes the general biological characteristics of sponges, their properties, uses and applications. Moreover, this study discusses a commercial fishery analysis of this natural resource in Cuba during the period 1970–2017, as well as the different characteristics of their natural populations subjected to commercial extraction. The applied techniques for aquaculture, harvest and postharvest processing are reviewed, including those procedures adapted from other countries or locally developed by Cuban fishermen. Finally, this study examines the challenges and perspectives of this productive activity with a long-term eco-sustainable approach.

Keywords: sponges, ecology, aquaculture, production, Cuba

1. Introduction

Sponges are very primitive pluricellular aquatic organisms that belong to *phylum Porifera*; they are sessile species and live adhered to different types of soft and hard substrates. *Porifera* are of diverse shapes and colours and have very varied sizes, from a few millimetres to more than two metres in height. They are found practically at all depths; non-polluted coastal zones and tropical reefs are especially rich in this type of species. The majority of sponges are marine, and only is the family *Spongillidae* known to live in fresh water. Until well into the eighteenth century, the origin of sponges was justified differently as nests of certain marine animals or plants. In 1786, John Ellis, an Englishman, observed retraction and expansion movements of the sponge pores, as well as water currents flowing through their bodies, so he classified them as animal organisms. Nevertheless, sponges lack some characteristics of typical animals, such as not having organs totally differentiated, which in a way could be considered equivalent to sets of specialised cells that pump water through their whole body to obtain oxygen, food, transport waste and reproductive products. Their reproduction is sexual and asexual; this last one is the most common as it occurs at any moment by gemmulation, formation of root-like extensions and fragmentation of the mother sponge. Sexual reproduction can be hermaphrodite with separate sexes in definite or temporal form. Oviparous and viviparous species have also been recorded; in both cases, the eggs or larvae are released through the exhaling water current [1–3].

Close to 9000 sponge species are estimated to exist in the world, but only do a few ones belong to the order *Keratosa* and the family *Spongiidae*, which have been traditionally utilised for commercial purposes [4, 3]. These kinds of sponges are characterised by having a corneal skeleton densely reticulated, free of spicules with a great capacity to retain water, elasticity and durability [5]. The main commercial value of natural sponges is given by their high capacity for water retention and because they can pump 1200 times their own volume per day. *Keratosa* and *Spongiidae* are able to retain up to 90% of particulate organic material, such as plankton, bacteria and even viruses suspended in water besides their resistance to acids and greater easiness for cleaning with respect to the artificial ones [6, 7].

The demand of natural sponges has been growing because of the market preference for natural products and the increase of their use in man's life, such as domestic, cosmetic, biomedicine, pharmaceutical, pottery, art industry, filter, cleaning and industrial purposes, among other uses. The commonly called 'bath sponges' of the family *Spongiidae* have a high price in the market although their offer has decreased due to natural reduction and impact of different natural and anthropogenic factors, among them, increase in frequency and intensity of extreme meteorological events, such as hurricanes, deriving from climate change or variability, pollution, disease incidence and overexploitation of natural populations [8]. The industrial and domestic sponges in Cuba have an international market that rises above 40 million USD annually [9].

Natural populations in sponge zones in the world, such as the Antillean region (Cuba, Bahamas and Florida), guarantee more than 50% of the world production. In Mexico, the Caribbean reefs (Isla Mujeres, QR) and the Gulf of Mexico have great species richness that includes the three classes that integrate *phylum Porifera: Calcarea*, *Hexactinellida* and *Demospongiae*. Nonetheless, no commercial exploitation of sponges exists in Mexico. Although none of the species of this taxonomic group are found protected by the Mexican norm NOM-059, natural sponge populations are located in protected natural areas, national marine parks or biosphere reserves under conservation legislation [3]. On the other hand, the production zones that stand out are the Mediterranean Sea (Syria, Turkey and Greece), the Adriatic Sea (Lebanon, Egypt, Tunisia and Italy) and the North Pacific Ocean (Philippine Sea, Carolina Islands and Marshall Islands) [4, 10].

The species from the Mediterranean are considered as those with the best quality and commercial value, of which those that stand out are the species *Spongia officinalis* or 'Fina' [in Spanish] (the best of all the commercial species; bath sponge), *Hippospongia communis* 'Común' [in Spanish] (of greater abundance; horse sponge), *Spongia zimoca* and *Spongia agaricina* or 'Oreja de Elefante' [in Spanish] (elephant ear; lamella) [4, 5].

In the Antillean region, the best commercial sponges have come from Cuba and the Bahamas Islands. Although several species have been reported in Cuba, four species have been the target for capture because of their abundance [1, 11–13]. Of the four species, three of them correspond to those commonly called 'machos' [males] from the genus *Spongia*: *Spongia barbara* (Duchassaing & Michelotti, 1864) called in Cuba 'macho fino' [fine male], *Spongia obscura* (Hyatt, 1877) or 'macho cueva' [cave male] and *Spongia graminea* (Hyatt, 1877) or 'macho guante' [glove male] and the one called 'hembra de ojo' [eye female] or Wool of the genus *Hippospongia* and *Hippospongia lachne* (Laubenfels, 1936), which is the one with the greatest commercial value in Cuba although other species usually show up in capture.

The presence of commercial sponges, as well as their fishing or recollection, has been reported in Cuba since the nineteenth century. During colony times, fishing boats from the Bahamas would reach the coasts of the Caribbean and Nuevitas (northeast of Cuba) to fish sponges with licence from Spanish authorities where more than 150 thousand dozen were fished by Cubans in 1867 [14]. Years later, sponge fishing was developed in southwestern Cuba with fishermen from Batabanó port, and because of their abundance, the two fishing zones in Cuba were established: (1) the northeast zone exploited by boats and fishermen from Caibarién where the Sabana-Camagüey Archipelago is located and (2) the Gulf of Batabanó, in southwest Cuba, exploited by boats from Batabanó fishing port. By 1886 offices in London and Paris were established to commercialise this product [15]. In 1930, the production went beyond 1 million dozen until 1939 and up to 1943 when a disease known as 'tizón' (blight), caused by the fungus *Spoingiophaga communis*, reduced the Cuban, Bahamas and Florida populations. Jointly with this situation, damage caused by a hurricane in 1944 led to a decrease in sponge production in the region to 16,000 dozen in 1947 [16].

This chapter discusses the principal studies and criteria related to commercial sponge fishery and aquaculture advances in Cuba, the main impacting factors that limit their abundance, and the challenges to increase aquaculture production of this important resource sustainably in the long term and with an ecosystem approach.

2. Analysis of the commercial sponge fishery in Cuba

Sponge fishery in Cuba has shown two extraction procedures, in accordance with the characteristics of the extraction zone, fishermen's age and regional traditions [17–19]: (1) by means of hooking implements for sponge recollection from auxiliary (small) boats that are towed by a sponsor, so fisherman immersion is not needed, or (2) by diving in apnoea for detaching or cutting the sponges from the closest part to the fixation substrate. Practically, no evolution in the fishing form has taken place throughout the years. The shallowness of the area where sponges inhabit has determined the fishing system that has followed the traditional method, using a glass bottom bucket and a stick with a double hook or trident to detach the sponge from the substrate (**Figure 1**).

Cuba reached an important commercial sponge production with an average of 166 t in the period from 1910 to 1919; 505 t for 1920–1929 and 391 t for 1930–1939 [20]. From 1939 to 1943, the fungus (blight) disease decimated the populations jointly with the hurricane at the end of 1944, generating lower production levels until 1947 [16, 21]. During the period after 1960, fishery activity was reorganised in Cuba; the fleet was modernised, which decreased the number of sponge boats and fishermen; fishing areas were divided into zones by territories, establishing

Figure 1.
Traditional technique for sponge capture or recollection in Cuba, sponge boat, auxiliary boat and glass bottom bucket. Photography: La Empresa Pesquera Industrial de Caibarien (EPICAI).

two fishing regions (**Figure 2**) in terms of abundance [22, 23]. Currently, commercial sponge fishery in Cuba is regulated by catch quota, and minimum legal sizes have been established for perimetric length: 35.6 cm for *H. lachne* ('Hembra de ojo'), 30.6 cm for *S. obscura* ('Macho cueva') and 20.8 cm for other species, such as *S. pertusa*, *S. barbara* and *S. graminea* [24].

Although current statistics have shown a tendency to increase sponge extractive activities since 1960, Cuba has not been able to reach the production levels previous to 1940. This tendency could have been due to a greater fishing effort. Almost all the fleets of Batabanó and Caibarién ports dedicated themselves to the capture of this resource and utilised boats type 'Balandro' and 'Goleta' with a crew from 14 to 16 fishermen. Before 1944 the fleets operated around 350 boats in the Gulf of Batabanó, which belonged to the Cuban ports of Batabanó, Coloma and Gerona [13, 14]. Production increased from 1960 with the proper fluctuations of a fishery that depended on different natural and human factors. Nonetheless, the average annual capture (40.15 ± 12.8 t) from 1960 to 2017 (58 years) did not go beyond 50 t (**Figure 3**).

Sponge fishery production decreased in southwest Cuba (Gulf of Batabanó) by fishing region from the beginning of the 2000 decade. Production reported by the enterprise PESCAHABANA (Batabanó) fell from 28.2 ± 3.1 t (2000–2004) to 19.6 ± 1.6 t (2013–2017). A similar pattern was registered for the northeast region (Sabana-Camagüey Archipelago). Production from the industrial fishery (EPICAI) decreased from 25.4 ± 1.6 t (2000–2004) to 14.1 ± 3.0 t (2013–2017). In the case of Caibarién, a greater stability was observed in sponge production during the period 1990–2009 (23 ± 3.8 t). Nevertheless, average capture from the period 2010–2017 was 19.1 ± 9.0 t with a maximum capture (>33 t) in 2010 and 2011, much higher than the historic average (23.5 t) from the period 1972–2017 (47 years). All these data suggested that overfishing occurred during 2010 and 2011 whose consequence was observed several years later with a lower extraction of 15 t, which affected national sponge production. The situation of this region got worse in 2017 (10.4 t) due to the impact of Hurricane 'Irma'.

2.1 Abundance by species and regions

Population density studies developed in the 2000 decade [13] showed a greater sample density in the region of the Sabana-Camagüey Archipelago (Caibarién) with respect to sample data for the region of the Gulf of Batabanó (**Figure 4**).

In both Cuban regions, northeast (Caibarién) and southwest (Batabanó), commercial extraction of the sponges locally known as 'Machos', which belong to the genus *Spongia sp*, was higher than those known as 'Hembras' (*Hippospongia lachne*).

Figure 2.
Distribution of the fishing areas according to zones of major commercial sponge abundance: northeast zone (Sabana-Camagüey Archipelago) and southwest zone (Gulf of Batabanó) Cuba.

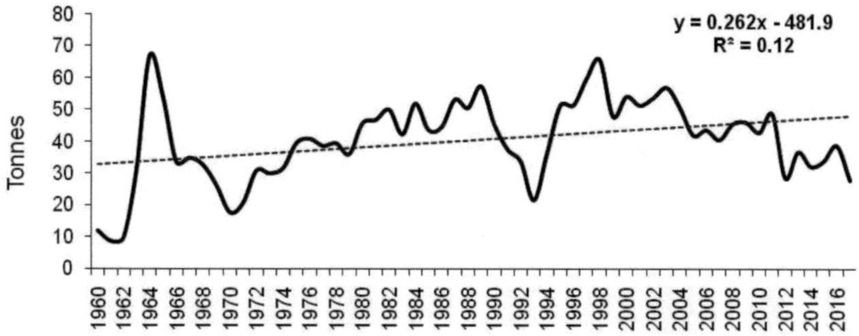

Figure 3.
Interannual variability of commercial sponge annual average extraction for the 1960–2017 period, in Cuba.

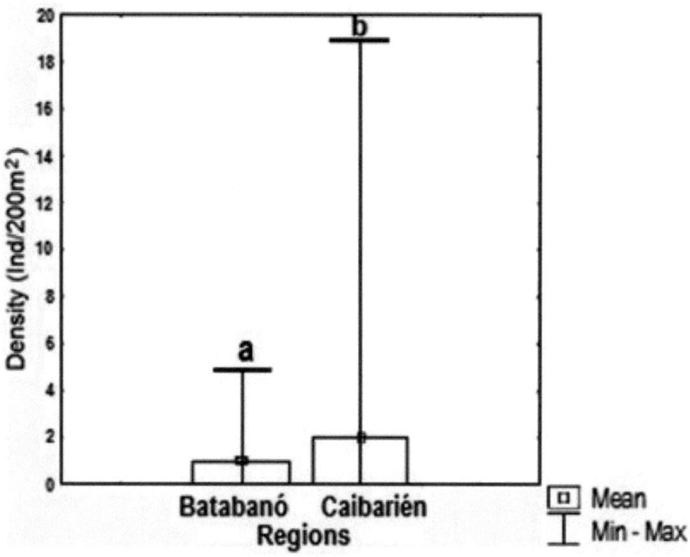

Figure 4.
Sponge density according to southwest (Batabanó) and northeast (Caibarién) regions in Cuba. Different letters indicate significant differences $p < 0.05$ *[13].*

On the other hand, regardless of the fishing effort applied in each region, a tendency to decrease was observed in sponge production, which was lower for *H. lachne* species both in extraction data and abundance parameters in both regions [13]. Species extraction of the genus *Spongia* sp. in the northcentral or northeast regions of Cuba was relatively constant during the 2000 decade. A different pattern was found for the same genus in the southern region (Gulf of Batabanó), showing a tendency to decrease, same as those of the species *H. lachne*. This situation was reflected on sample density by species (**Figures 5** and **6**).

In subsequent studies developed in a protected northeast zone of the Sabana-Camagüey Archipelago during 2013 [25], which is under the Special Regime of Use and Protection and in which commercial sponge extraction has not been performed, an average density of 0.457 ind/m^2 was recorded (4570/ind/ha), estimating a potential precautionary capture of 1300 specimens (30% of the total) per hectare per year [25].

Density data obtained were by far superior to those reported by Blanco and Formoso [13] for the extraction zones of the Sabana-Camagüey Archipelago. Furthermore, they were superior to those reported by these same authors for the

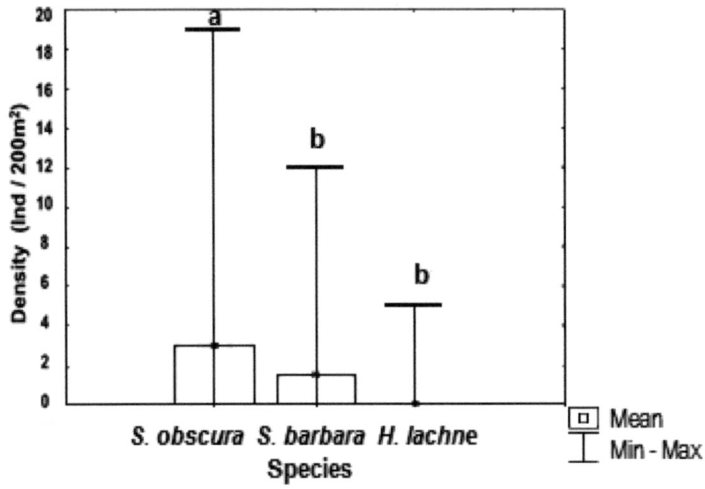

Figure 5.
Commercial sponge density by species on the natural banks of northeast Cuba (Sabana-Camagüey Archipelago). Different letters indicate significant differences p < 0.05 [13].

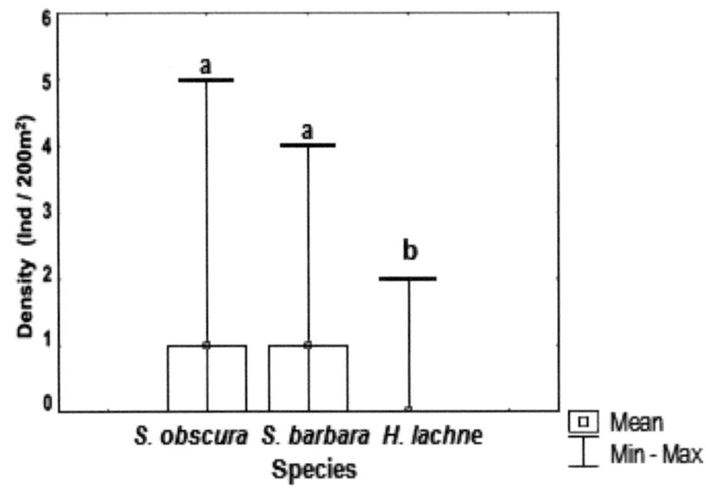

Figure 6.
Commercial sponge density by species found on natural banks of southwestern Cuba (Gulf of Batabanó). Different letters indicate significant differences p < 0.05 [13].

Gulf of Batabanó in the 2000 decade, which evidenced, among other causes, that fishery was also an impact factor.

What is more transcendental data from recent studies is that the 'Hembra' sponge *H. lachne* showed greater abundance in the zone assessed from the protected area in 2013. Just because of its greater economic value, it is one of the first species to decrease in abundance, and it also evidenced that a decrease in production is expected due to commercial exploitation and improvement of extractive methods [25].

3. Main impact factors in sponge populations in Cuba

As previously mentioned, one of the main causes of abrupt decrease in sponge populations in Cuba was related to the disease known locally as 'Tizón' (smut or blight) caused by the fungus *Spoingiophaga communis* during the period 1939–1945 [17, 20, 21]. After 1945, no significant outbreaks of this disease were reported. Because of the capture decrease in the Gulf of Batabanó, studies were performed in 2005, but no proofs of the existence of pathogenic organisms (fungi and bacteria), which could cause sponge death and subsequent decrease in production, were detected that year [26].

Solar radiation, illumination and temperature are factors that regulate sponge distribution, colonisation and success in their natural reproductive processes. Although they can withstand extreme temperature (10–36°C) values in short periods, the optimum values for their sexual proliferation is from 23 to 29°C [27, 28]. In Cuba the southern sponge zones of the Gulf of Batabanó showed water temperature average of 28.03°C, while in the northern zone of Sabana-Camagüey Archipelago, average temperature was 27.33°C in Sabana and 28.32°C in Camagüey [29].

Even though these values are permissible for commercial sponges in Cuba, high temperatures (>30°C) can also favour the proliferation of bacteria and fungi. In coastal water bodies and bays in the inner part of the Camagüey Archipelago, extreme maximum temperatures up to 35°C could occur due to shallowness and limited water renovation [29]. Because of the shallowness from 3 to 7 m in the Gulf of Batabanó and from 2 to 8 m in the Sabana-Camagüey Archipelago, in which the greatest abundance of Cuban commercial sponges inhabit, they are very vulnerable to natural physical impacts.

Blanco [23] pointed out hurricanes as a cause of impact on sponge populations, above all on those that inhabited the Gulf of Batabanó due to a greater frequency and intensity of cyclonic disturbances after 1996. Hurricanes generate strong currents and surge of great height and intensity that provoke sediment in suspension besides the fracture and dragging of fragments or complete organisms. It occurs to sponges themselves due to their sessile condition that makes it impossible for them to escape from the energetic movement that occurs in waters, which makes the effect greater on the genera *Spongia* and *Hippospongia* because they are very susceptible due to their high tissue density [30–32].

The increase of anthropogenic activities, such as tourism development in keys and islands, above all in the Sabana-Camagüey Archipelago, adds contamination and increase in water turbidity; dragging and landfill for construction and repairing roads that link the coast of Cuba to these keys have led to periodical turbidity events that have affected seawater quality [33]. The excess of small particle solids suspended in the water column has caused clogging of the inhaling pores in commercial and noncommercial species, more so in those that have fine pores, causing them inadequate development, including death [1, 26]. The increase of siltation due to coastal erosion has been another impact additional to hurricanes, which has been derived from logging bordering mangroves, maritime construction and increase of average seawater level, as it has occurred in several coastal segments in the southwest region of the Gulf of Batabanó.

On the contrary, organic contamination at intermediate degrees seemed to have caused certain stimulation to sponge development and diversification, but it also reduced species diversity in reefs dramatically and, in extreme cases, has a greater decrease of their biomass [34]. Contamination has also brought as a consequence the disappearance of marine grass rich in commercial sponges and its substitution for muddy bottoms with turbid water loaded with sediments that do not favour *Porifera*. This situation has occurred in wide zones of the Sabana-Camagüey Archipelago [34].

Finally, fishing activity itself could constitute an additional impact when resource exploitation goes beyond its recovery capacity since uncontrolled extraction levels lead to overfishing patterns. Blanco [23] points out a tendency of sponges to decrease, above all, the species *Hippospongia lachne*, associated to the high exploitation rate to which it has been subjected for years, among other factors, due to its high commercial value.

4. Sponge culture in Cuba

The development of sustainable and economically viable fishery production alternatives, such as sponge culture, constitutes an additional contribution to environment sustainability. It is a working alternative for fishermen to create new community employment sources and generate income of foreign currency besides the need of moving from a predatory recollection activity to a productive aquaculture work, as a step in economic and cultural fishery development in the country [8].

Sponge culture offers a safe and predictable production of a superior quality product to that offered by natural capture besides its elevated price according to the market, quality and species. Besides the easiness of their collection in their natural environment because they are sessile organisms that are generally found in shallow waters, they do not need additional food to that filtered from their environment. This is the reason why its culture requires low investment cost and availability to schedule a tiered harvest. Moreover, its culture reduces fishing pressure on sponges in their natural medium, constituting a sustainable repopulation alternative to increase natural banks surrounding the aquaculture farms because of their larval contribution to the environment [35].

Initial sponge culture in Cuba goes back to several decades. A variance of sponge culture suspended in vertical lines was tested in Cuba in 1965 and described by García del Barco [36, 37] in a sponge culture handbook. The method of vertical suspended lines allowed using a greater area vertically taking advantage of the zone in a greater depth and avoiding being affected by surge as it occurs in lower zones where they traditionally inhabit.

Complete experimental cycles included sponge collection from their natural environment, seeding, harvesting and reseeding from seeds obtained from the same culture, cleaning process and commercialisation. Aquaculture procedures were performed with the assessment of scientific institutions, such as Centro de Investigaciones Pesqueras de Cuba [38, 39].

Although sponge culture was not consolidated to a commercial level, important conclusions were obtained from these studies:

- Cultured sponges showed less osculation density and diameter, increasing solid surface and weight per volume unit.
- They showed less mechanical damage during recollection.
- Cultured sponges were harvested in total absence of foreign materials.
- They showed spherical shapes which reduced process expenses and wastes.
- Cultured sponges reached a similar or greater size to those in their natural environment, in equal period, but with better and more rounded shape.
- 'Seeds' for a nondependant aquaculture could be obtained from their natural environment if not harvesting a part of the cultured sponges and allowing them to naturally grow for about 3 years to get a 'mother sponge'.

The technical and scientific knowledge and field experiences derived from these experiments allowed editing a handbook of work procedures and operations for small sponge farms attended by the same extractive fishery crew [40]. Research and development has continued, and two culture methods have been tested during the last decade, which are briefly described below.

4.1 Free sponge method

An experimental farm was projected by the Centro de Estudios y Servicios Ambientales (CESAM, its abbreviation in Spanish for Centre for Environmental Studies and Services) of Villa Clara, Cuba. It was sponsored by funding partners of the United Nations Development Programme for Global Environmental Finance (Small Donations GEF-PNUD). The sponge farm was located in a marine zone in the surroundings of the town Carahatas (Sabana-Camagüey Archipelago) north-central coast of Cuba. One-hectare culture fences were built and installed in the sea. Metallic poles were buried in the seabed as basic support and plastic mesh cove to restrict access to predators. The 'free' sponge method was used in those subdivided 1-ha lots, planting a density of 1 sponge/4 m^2 (**Figure 7**).

Starting from the contribution of the project GEF/PNUD/'Protección de la biodiversidad en tres sectores productivos del Archipiélago Sabana-Camagüey' [Biodiversity protection in three productive sectors of the Sabana-Camagüey Archipelago], fishermen from the Caibarien Basic Enterprise Unit (EPICAI) built a farm in a northeast shallow marine zone with the advice from Centro de Investigaciones Pesqueras.

Recollected sponges were cut in 4–5 cm^3pieces named as 'propagules' that were used for 'seeding' and deposited in the substrate (approximately 2500 seeds/ha), at the mercy of currents and other natural water dynamics, until they reached a commercial size. A total of 12 ha seeded were obtained, which should provide 1 t of sponge in a year at a quote of more than $15,000 USD in the world market [41].

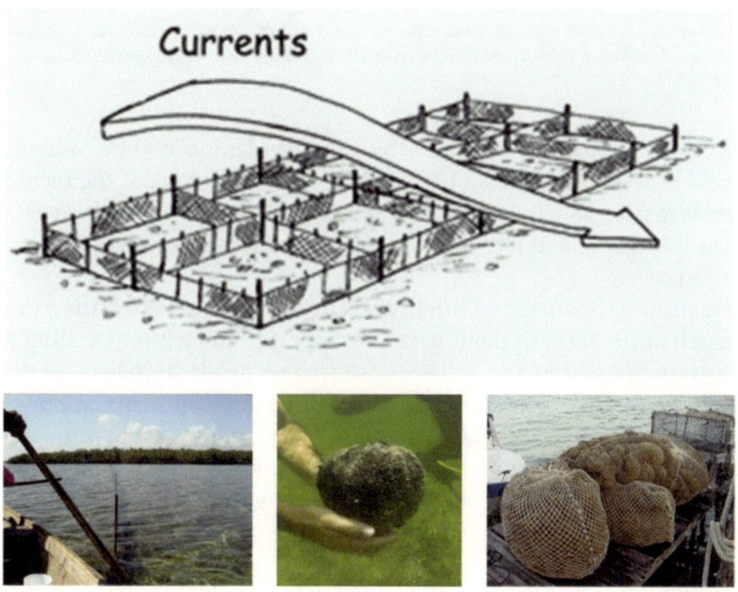

Figure 7.
Experimental sponge farm in Carahatas, Sabana-Camagüey Archipelago, Cuba. Free sponge culture in lots. Graphic art and photography: [41].

4.2 Trials and tendales method

This method uses rope 'tendales' in a horizontal pattern, which is commonly identified in aquaculture as 'suspended long-line' method. Briefly, metallic poles are buried in the seabed as support for nylon-braided rope (long lines 1/4″), elevated 20–30 cm off-bottom. Two long lines support horizontal several tendales of nylon monofilament (150 lb) for sponge suspended aquaculture. Mother sponges are cut in propagues (5 to 8 cm^3) to obtain sponges seeds. Propagues are tied to tendales in a collar-shape pattern using monofilament nylon lines (50 lb). In this way, sponge 'seeds' hang vertically to horizontal tendales with a separation of 30 cm between each one, during all the grow-out period (**Figure 8A**). Alternatively, sponge seeds can be put directly in the nylon tendales (**Figure 8B**).

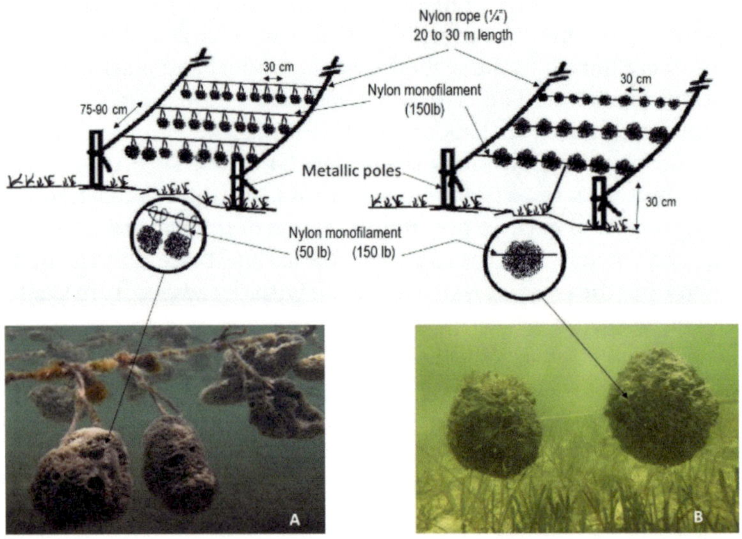

Figure 8.
Suspended line culture. System designed for the experimental farm in Caibarién, Sabana-Camagüey Archipelago, Cuba. Graphic art: M.A. Avilés-Quevedo. Photography: Empresa Pesquera Industrial de Caibarien (EPICAI).

After a grow-out from 15 to 18 months, 80% of planted sponges were obtained with acceptable commercial size (18–23 cm in diameter). Part of the recollection of this farm was used as 'mother sponge' to obtain new lots of 'seeds' for a second project with 130 suspended lines (trails), each one with 33 sponges for a total of 4290 cultured sponges [42].

All these projects, efforts and intentions to boost sponge cultivation in Cuba have remained at the stage of demonstrative experiments without scaling up to allow expansion to a systematic and eco-sustainable production level with an economic profitable income. The causes of this limited development have been related rather than beyond the indisputable potential of marine waters to human factors related to the will of introducing, developing and consolidating sponge culture, which could promote a regional socioeconomic progress.

5. Challenges

Gradual reduction of natural sponge banks at national and global levels has been evident, and that risk situation could get worse due to the problems deriving from climate change. Sponge culture, besides being a sustainable production, constitutes an alternative in foreign currency with commercialisation prices according to Cuban commercial species from $4 to $74 USD/kg, depending on their quality classification.

The main challenges to develop and generalise sponge culture in Cuba are:

1. Link and implicate fishery enterprises and coastal communities to develop sponge culture projects.

2. Assess and select ideal sites for priority species, according to value and abundance of the natural resources, to implement a viable economical and eco-sustainable aquaculture.

3. Apply a differential price and payment policy to fishermen, according to natural and cultured sponges. It is essential although clearly established policies exist for the development of marine aquaculture in Cuba.

4. In other terms, cultured sponges should have more attractive prices to motivate their introduction and boost technologic development and generalisation or the activity.

5. The economical-environmental feasibility that fishermen themselves combine natural sponge extraction with aquaculture production may not be viable in practice due to their extractive tradition, timing annual fishery operations and compliance demand for official production plans or goals, among other subjective factors.

6. Facing the decrease in sponge capture and abundance, it shall be essential to reduce fishing effort on natural populations, diverting fleet and fishermen that are currently dedicated to sponge extraction towards aquaculture production.

7. Those challenges will imply economic and logistic support from state institutions until the first results have been reached, and after that first goal, a second step of continuity will be necessary to improve and continuously enhance this productive activity.

Acknowledgements

This study was financed by Public Sectorial Research Fund for Education of México, Basic Science project Conacyt No. 258282 and R&D project Proinnova Conacyt No. 241777, under the academic responsibility of JMMS. Authors are grateful to Centro de Investigaciones Pesqueras (CIP, Cuba), Centro de Investigaciones Biológicas del Noroeste, S.C. (CIBNOR, México), Antonio Grovas Hernández of Grupo Empresarial de la Industria Alimentaria (GEIA, Cuba) who provided updated sponge extraction data, and Diana Fisher for English edition.

Author details

Betanzos-Vega Abel[1], Mazón-Suástegui José Manuel[2*], Formoso-García Mario[1] and Avilés-Quevedo María Araceli[3]

1 Centro de Investigaciones Pesqueras (CIP), La Habana, Cuba

2 Centro de Investigaciones Biológicas del Noroeste S.C., La Paz, BCS, México

3 Freelance Services for R & D in Marine Aquaculture, La Paz, BCS, México

*Address all correspondence to: jmazon04@cibnor.mx

References

[1] Alcolado PM. Las Esponjas. Editorial Científico-Técnico, Ministerio de Cultura; 1986. 56 p

[2] Gómez P, Carballo JL, Vázquez LE, Cruz JA. New records for the sponge fauna (Porifera : Demospongiae) of the Pacific coast of Mexico (Eastern Pacific Ocean). Proceedings of Biological Society of Washington. 2002;**115**:223-237

[3] Torruco-Gómez A, González-Solís A. Diversidad Faunística, invertebrados Esponjas y su importancia. In: Durán, Mendez, editors. Biodiversidad 496. 2010. pp. 202-203

[4] Espinosa L, González D. Sinopsis de las Esponjas Comerciales en Cuba. In: XV Foro de Ciencia y Técnica. CIP, MIP; 2002. p. 14

[5] Pronzato R, Gaino E. La Malattia della spugne commerciales. Considerazion estorico-economiche. In: Rovil Mus. Int. Biol. Univ. Genova. 1991. pp. 54-55

[6] Carballo JL, Gómez P, Cruz-Barraza JA. Biodiversidad de Porífera en México. Revista Mexicana de Biodiversidad. 2014;**85**(Suppl):143-153

[7] Nava H, Figueroa Camacho AG, Haro Medrano CA, Villegas Sánchez BM. Esponjas marinas: Importancia ecológica en los ecosistemas arrecifales. CONABIO. Biodiversitas. 2015;**123**:12-16

[8] Betanzos-Vega A, Millares N. Cultivo de esponjas en ambiente marino. Opción de desarrollo pesquero sostenible en Cuba. Asociación Cubana de Producción Animal. 2013;**2**:20-21

[9] Duckworth A, Wolff C, Evans-Illidge E. Distribution and Abundance of Bath Sponges in Torres Strait. Townsville, Queensland: Australian Institute of Marine Science; 2005. 23 p

[10] Espinosa L. Incremento de la Eficiencia de las Pesquerías de Esponjas Comerciales en Cuba. Informe de Resultado; Centro de Investigaciones Pesqueras. Ministerio de la Pesca. 2005. 11 p

[11] Alcolado PM. General features of Cuban sponge communities. In: Rützler K, editor. New Perspectives in Sponge Biology. Washington, DC: Smithsonian Institution Press; 1990. pp. 51-57

[12] Alcolado PM, Grovas-Hernández A, Marcos Z. General comments on species inventory, fisheries, culture and some community features of Porifera in Cuba. Bollettino dei Musei e Degli Istituti Biologici dell'Università di Genova. 2004;(68):175-186

[13] Blanco JC, Formoso M. Estimado de la abundancia de esponjas comerciales en áreas correspondientes a las plataformas suroccidental (Golfo de Batabanó) y norcentral (Archipiélago Sabana-Camagüey) de Cuba. Revista de Investigación. 2009;**30**(2):99-106

[14] Sánchez-Roig M. La riqueza esponjícola cubana. Mar y Pesca. 1957;**1**(3):28-32

[15] Arrazcaeta R, Chong F. La epesca y el comercio marítimo en Batabanó. Mar y Pesca. 1983;**216**:16-21

[16] Radcliffe L. The sponge industry from an economic point of view. Proceedings of the Gulf and Caribbean Fisheries Institute. 1949;(1):43-46

[17] Cardona R. Informe sobre la especie esponja; MIP-Cuba1983. p. 34

[18] Páez J. El recurso esponja de la Plataforma Cubana, I: Síntesis histórica, biología y ecología. Revista de Investigaciones Marinas. 1990;**11**(1):11-25

[19] Witzell WN. The origin of the Florida sponge fishery. Marine Fisheries Review. 1998;**60**(1):27-32

[20] Stuart AH. World Trade in Sponge. Serie 82. U.S. Dept. Comm. Industrial; 1948. 95 p

[21] Kahn PA. The legislative situation on sponges. Proceedings of the Gulf and Caribbean Fisheries Institute. 1950:16-34

[22] Baisre JA. La pesca marítima en Cuba. Editorial CientíficoTécnica Ministerio de Cultura; 2004. 338 p

[23] Blanco JC. Esponjas comerciales de Cuba. Escenario actual. Tesis presentada en opción del grado de Maestro en Ciencias, Centro de Investigaciones Marinas, Universidad de La Habana, Cuba. 2007. 74 p

[24] Grovas-Hernández AJ, Oliva-Mieres D. Análisis de las pesquerías de esponjas de la Asociación PESCABAT. Mar y Pesca. 1999:319-330

[25] Lopeztegui A. Evaluación de la abundancia de esponjas comerciales en una zona protegida del Archipiélago Sabana-Camagüey. La Habana, Cuba: Archivo del Centro de Investigaciones Pesqueras; 2013. 5 p

[26] Cruz-Quintana Y. Informe sobre muestreo por mortalidades de esponjas comerciales en la región del Golfo de Batabanó. Archivos delCIP; MIP-Cuba. 2005. 3 p

[27] Storr FJ. Ecology of the Gulf of Mexico Commercial Sponges and its Relation to the Fishery. Special Scientific Report Fisheries. Vol. 4661964. 73 p

[28] Alcolado PM. Estructura ecológica de las comunidades de esponjas en Punta del Este, Cuba. Rep. Invest. Inst. Oceanol. Acad. Cien. Cub. 1985;**38**:1-65

[29] Fernández-Vila LJ, López-García DO, Ramírez-Stout. Características termohalinas de las aguas de la plataforma cubana. Memorias del VI Taller Internacional CONyMA, 2010. CD-ROM. ISBN 978-959-300-008-6. 18 p

[30] Salazar-Vallejo S. Huracanes y biodiversidad costera tropical. Revista de Biología Tropical. 2002;**50**(2). 10 p

[31] Wuff J. Effects of hurricane on survival and orientation of large erect coral reef sponges. Coral Reefs. 1995;**14**:55-61

[32] Wulff J. Rapid diversity and abundance decline in Caribbean coral reef sponge community. Biological Conservation. 2006;**127**:167-176

[33] Betanzos-Vega A, Capetillo N, Lopeztegui A, Martínez-Daranas B. Variación espacio-temporal de la turbidez y calidad en cuerpos de agua marina de uso pesquero, región norcentral de Cuba, 2008-2010. Serie Oceanológica. 2013;**12**:24-35

[34] Alcolado PM. Reading the code of coral reef sponge community composition and structure for environmental biomonitoring: Some experiences from Cuba. In: Custódio MR, Lobo-Hajdu G, Hajdu E, Muricy G, editors. Porifera Research: Biodiversity, Innovation and Sustaniability. Rio de Janeiro: Museu Nacional; 2007. pp. 3-10

[35] Yi Q, Wei Z, Hua L, Xingju Y, Meifang J. Cultivation of marine sponges. Chinese Journal of Oceanology and Limnology. 2005;**23**(2):194-198

[36] García del Barco F. Manual de Esponjicultura. La Habana, Cuba: MIP, CIP; 1972. 10 p

[37] Ubeda L. Sembrando esponjas en el Golfo de Batabanó. Mar y Pesca. 1980;**183**:32-35

[38] Artiaga L. Análisis de los resultados de la cosecha de un parque de cultivo de

esponjas en el CPI Batabanó. Trabajo de Diploma, Inst. de la Pesca "Andrés González Lines"; La Habana, Cuba. 1985

[39] Gutiérrez F. Informe Sobre el Cultivo de Esponjas en la Empresa Pesquera de Matanzas. 2 p

[40] Grovas-Hernández AJ. Manual de procedimientos operacionales de trabajo para las minigranjas esponjícolas atendidas por tripulaciones extractivas; Ministerio de la Industria Alimentaria, Dirección de Pesca. 2011. 34 p

[41] Quirós A, Perdomo ME, Rodríguez EL. Silverio. Cultivo de esponjas marinas. Manual de buenas prácticas. Resultado del Proyecto CUB/OP/2/07/09 "Uso Sostenible de los Recursos Naturales en la Comunidad Costera de Carahatas"; CESAM-CITMA, Villa Clara, Cuba. 2009. 15 p

[42] Betanzos-Vega A, Valle S. La subsistencia de los pescadores: Alternativas y prácticas sostenibles. In: Menéndez-Carrera L, Arellano-Acosta M, Alcolado PM, editors. Tendremos Desarrollo socioeconómico Sin conservación de la Biodiversidad? Experiencias del Proyecto Sabana-Camagüey en Paisajes Productivos; La Habana: Editorial AMA. 2015. pp. 174-185. ISBN: 978-959-300-105-2. Impresos dominicanos, S.R.L. 223 p

Chapter 7

Spectral Discrimination of Live and Bleached Corals: A Case Study on *Turbinaria peltata* (Esper, 1794) Using Field Spectroscopy

Nandini Ray Chaudhury, Ashwin Gujrati, T.V.R. Murthy and C.H. Satyanarayana

Abstract

Scleractinian corals represent the foundation species of reef ecosystems. Bleaching is a physiological, cellular response to environmental stresses wherein marine invertebrates including corals expel their endosymbiont, unicellular microalgae or zooxanthellae from their host tissues. Field spectroscopy helps to characterize the health of corals in terms of reflectance spectra or spectral signatures, i.e. reflected light as a function of wavelength. This chapter reports a case study on spectral discrimination of in situ hyperspectral signatures of live, apparently healthy and bleached corals collected from a single colony of *Turbinaria peltata* (Esper, 1794) sampled from Laku Point reef in Gujarat coast of India. Derivative analyses on the in situ reflectance data identify five narrow windows in the visible light region (green and red light regions) to spectrally discriminate live and bleached coral polyps of the *T. peltata* species. This study highlights the potential of field spectroscopy in characterizing coral health in situ through non-invasive sampling.

Keywords: coral, coral bleaching, *Turbinaria peltata*, spectral signature, derivative analysis

1. Introduction

Corals are foundation species of coral reef ecosystems. Hermatypic or reef-building corals are exclusively polypoid, marine organisms which belong to the taxonomic order: Scleractinia of Class Anthozoa and Phylum Cnidaria. Scleractinian corals are different from other Anthozoans like soft corals and sea anemones thanks to their continuous hard calcium carbonate crystal exoskeleton. Accordingly, corals can be ecologically divided, but cannot be systematically classified, into reef-building (hermatypic) and non-reef-building (ahermatypic) corals [1]. Hermatypic corals commonly contain millions of endosymbiotic, unicellular, dinoflagellate algae or zooxanthellae. Hence, they are also known as zooxanthellate corals. On the other hand, ahermatypic corals mostly lack zooxanthellae [1].

IntechOpen

There is a mutualistic symbiosis between the host coral polyps and the endosymbiont, unicellular, microalgae named zooxanthellae. Zooxanthellae owe their common name due to their yellow-brown colour and belong to the genus *Symbiodinium* sp. to eight lineages (clades A–H) based on phylogenetic classification [2]. Zooxanthellae photosynthesize and help host corals to meet their energy requirements from photosynthetic products, while the host provides them intracellular space and essential nutrients like nitrogen, inorganic carbon. The photosynthetic pigments within zooxanthellae along with the host tissue and calcium carbonate exoskeleton pigments give corals their essential colours. Bleaching is one of the common expressions of physiological response to environmental stresses wherein corals or any other zooxanthellate marine invertebrate organisms expel the zooxanthellae from their host tissues. The expulsion of zooxanthellae and the resultant reduction of zooxanthellae pigment concentration per host cell lead to visible paling/fading or whitening of the host organism. This process is known as bleaching. Bleaching results in varying levels of mortality of the host organism depending on the severity of stress. Thermal stress is considered as the principal cause for coral and other zooxanthellate invertebrate bleaching, while other environmental factors can also cause bleaching either independently or synergistically with thermal stress [3]. These abiotic factors include exposure to supra-optimal irradiances of visible radiation, exposure to ultraviolet (UV) radiation, low-temperature thermal stress, salinity changes, sedimentation and desiccation due to low-tidal exposure [3, 4]. Thermal stress alone or in combination with exposure to high irradiances of both visible and UV radiation leads to photoinhibition of photosynthesis in zooxanthellae as these stress conditions damage the photosystem II reaction centres of the zooxanthellae [3]. Bleaching results in varying levels of mortality of the host organism depending on the severity of stress [5].

Field level detection of coral bleaching often gets complicated by various physiological (e.g. in case of coral diseases) and physical/environmental factors (like turbid water) at colony level. Semi-quantitative data provided by refined colour scales like the Coral Health Chart developed by the CoralWatch programme [6, 7] are considered useful for a synoptic description of colony-scale bleaching status during rapid field surveys. The Coral Health Chart serves as a unique utility tool to document the colour transformation of corals over six colour stages during a bleaching condition when the coral loses its own colour saturation and the whiteness/brightness increases due to the loss of zooxanthellae and their pigments [6]. The chart helps in identifying and monitoring the coral health condition based on corals' apparent colour and thus provides a quick, inexpensive and non-invasive way of spot sampling. However, the utility of the Coral Health Chart gets limited if one wishes to understand the pigment level changes that happen in a coral during a bleaching condition.

Field spectroscopy offers an essential support as a non-invasive, proximal remote sensing sampling technique for hyperspectral characterization of sessile, benthic substrates like corals. Field spectroscopy involves the study of interrelationships between the spectral characteristics of objects and their biophysical attributes in the field environment [8]. Spectroscopy involves the collection and characterization of continuous spectra acquired in laboratory (reflectance), in situ with portable and waterproof radiometers (radiance reflectance) and even by remote sensing (remote sensing reflectance). The spectra are analysed in terms of intensities and shapes according to the absorption features characteristic of pigment compounds of the targets.

Few studies have demonstrated the spectral differences that exist between healthy and bleached corals based on in situ hyperspectral signatures [5]. In one of the pioneering attempts, Holden and LeDrew [9, 10] demonstrated that hyperspectral signature of bleached corals (in 400–700 nm) is significantly different than that of healthy corals sampled from a protected lagoon in Fiji in the South Pacific and from a beach location in Indonesia. They used clustering and ordination analyses along with derivative spectroscopy to discriminate the reflectance spectra of healthy and bleached corals. Clark et al. [11] further investigated the spectral distinction of live and dead corals (in 400–750 nm) at various stages of mortality and algal colonization as sampled from Rangiroa atoll, French Polynesia, soon after the mass coral bleaching of 1998. Their field experiment revealed that recently, dead corals had a relatively pronounced peak around 550–600 nm, and the degree of sharpness or peakedness around 550 nm was used to discriminate live coral from the dead with an accuracy of 88%. However, they also pointed out that efficacy of spectral discriminators does vary at different water depths due to water column attenuation. Another study [12] detected negative shift of the red edge in the reflectance spectra of experimentally stressed and naturally bleached corals. These studies recommended derivative spectroscopy as a promising tool for spectral discrimination of bleached and healthy corals through proximal remote sensing and even later from airborne or space-borne remote sensing data. Accordingly, comparison of in situ spectral characteristics of healthy and bleached corals becomes a prerequisite to develop an understanding on the spectral behaviour of bleached corals.

In this direction, a case study was carried out on spectral discrimination of in situ hyperspectral signatures of live and bleached corals collected from a single colony of *Turbinaria peltata* (Esper, 1794) sampled from Laku Point reef of Poshitra in Gulf of Kachchh during March 2011.

2. Sampling site for field experiment

The sampling site for this particular study, i.e. Laku Point reef (**Figure 1**), is located in the coastal village of Poshitra situated in the Okhamandal region of Gulf of Kachchh in Devbhoomi Dwarka district in Gujarat state of India [5]. Coral reefs, mangroves and rocky shores are major habitats of this site [13]. This is a narrow fringing reef connected to the mainland coast. The coastline is indented with small embayments, 1–2 km long and 0.5–1 km wide [14]. This site is marked with 100-m-wide eulittoral fringing reefs having high coral diversity [15]. Coral colonies grow in shallow, rock pools in the upper eulittoral zone. The rocky pools are covered by barnacles and oysters and produce a rugged topography. These rock pools are found in vertical tiers and exhibit variation to coral distribution and diversity according to tidal exposure. Laku Point site represents prominent biokarst landscape with vertical pinnacles or coastal lapies and pits or pools on the beach rock surface similar to landscapes reported from the Dwarka coast [16]. The common coral genera reported earlier from this site include *Turbinaria*, *Montipora*, *Favia*, *Favites*, *Porites*, *Goniopora* and *Goniastrea* covering 45% of the reef area [14].

3. Field experiment and data processing

The field experiment for the above-mentioned site was meticulously planned with reference to the Survey of India (SOI) tide table information considering Okha (22°58′ N, 70°27′E) as the reference tidal station. Laku Point reef in Poshitra was sampled during 20–24 March 2011. The maximum negative tide was −0.09 m on 22 March 2011 [17]. The equinoctial spring tide windows in a year offer suitable conditions for passive, proximal sensing of corals with minimal water column as the

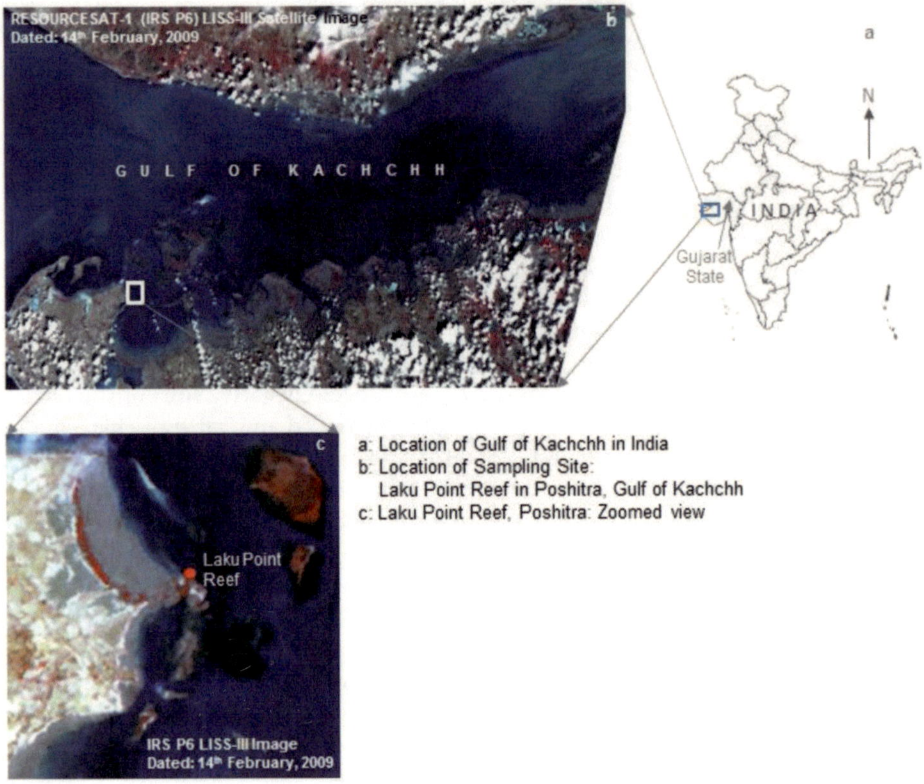

Figure 1.
Location of the sampling site: Laku Point Reef, India ((a) location of gulf of Kachchh in India, (b) location of Laku Point reef in gulf of Kachchh, (c) Laku Point reef, zoomed view from Indian remote sensing satellite).

low tide exposures of reefs coincide with early hours of local day time (i.e. 09:00–11:00 hours) with clear sky conditions [5, 18, 19]. Reflectance spectra of the sampled hard coral were collected with analytical spectral devices (ASD) FieldSpec®3 spectroradiometer having a spectral range of 350—2500 nm and spectral resolution of 3 nm (at 700 nm) and 10 nm (at 1400 and 2100 nm) [20]. The sampling interval is 1.4 nm for 350–1000 nm spectral region and 2 nm for 1000–2500 nm. The visible and near-infrared (VNIR) spectral region (350–1000 nm) in this spectrometer is configured with 512 element silicon photodiode array, while SWIR1 (shortwave infrared: 1000–1830 nm) and SWIR2 (1830–2500 nm) spectral regions are configured with indium gallium arsenide (InGaAs) detectors. The fibre-optic probe has a field of view (FOV) of 25° full conical angle. The spectra were measured holding the optical probe at a minimum height of 30 cm above the target with a nadir view, and due care was taken to ensure that the target diameter was always greater than 15 cm. The field spectroradiometer was calibrated with reference to a Spectralon white plate (or 100% white reference standard), and thereafter multiple spectra were recorded for different sample surfaces, i.e. apparently healthy, partially bleached and bleached (**Figure 2**).

Turbinaria peltata (Esper, 1794) is a representative of *Dendrophylliidae* family (Gray, 1847) of scleractinian corals. This species occurs as flat, plate surface colonies in grey to brown colour. *T. peltata* grows in the rocky foreshores and shallow reef slope zones even in turbid water [15, 21].

The *Turbinaria peltata* (Esper, 1794) coral colony sampled (**Figure 3A** and **B**) had three representative surfaces: (i) apparently healthy, live coral cover, (ii) partially bleached surface and (iii) bleached surface. This distinction was made in field with

Figure 2.
Field spectroradiometer and in situ hyperspectral data collection at Laku Point reef site, India [5].

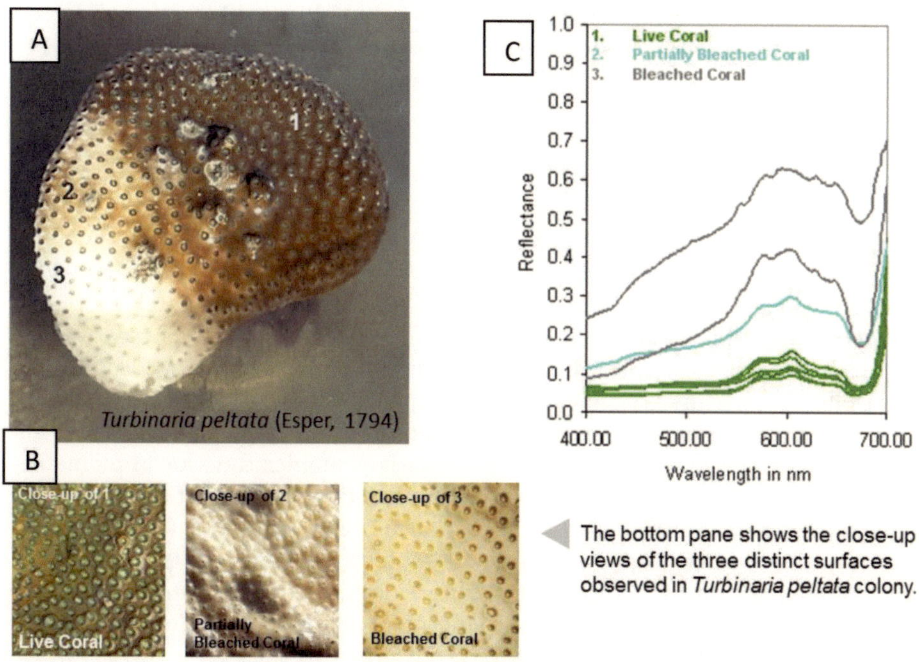

Figure 3.
In situ reflectance spectra of coral species: Turbinaria peltata *in live, partially bleached and bleached condition (the MRS in each class represents the number of sample spectral measurements for three different surface conditions in the coral colony, e.g. (i) live coral = 5, (ii) partially bleached = 1 and (iii) bleached = 2) [5].*

reference to colour differences perceivable to human eye during field sampling. In situ reflectance spectra were collected from each of these surfaces. For each sample surface, a minimum of 30 reflectance spectra were logged with the help of spectra acquisition software: RS3. Field photographs of the sample surfaces were also taken with a digital camera and sequentially numbered. Data logging was completed within a 15-minute period for each sample surface. The field spectra were subsequently processed with the help of ViewSpec Pro software (version 5.6).

The mean representative spectrum ((MRS) and n = 30) of these three surfaces was first plotted (**Figure 4**) for the visible region, i.e. 400–700 nm range for visual comparison. It was found that beyond 715–1350 nm, bleached coral spectra closely

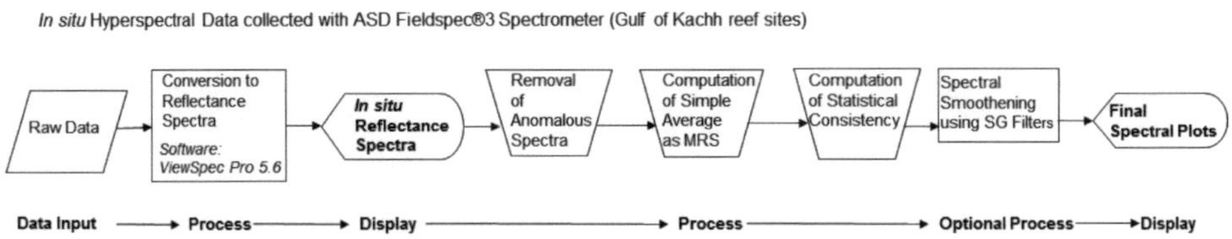

Figure 4.
Flow chart showing in situ hyperspectral data processing [5].

follow the trend of the live corals with only local shoulders and troughs getting vertically pronounced [18].

For each sample surfaces, in situ reflectance spectra (n = 30) were first plotted for visual appreciation and data editing. Anomalous spectra matching neither in magnitude nor with respect to the shape of the rest of the spectra were first manually removed. The remaining spectra were arithmetically averaged to obtain the simple average spectra or the mean representative spectrum (**Figure 4**) for the corresponding sample surfaces. The consistency of the spectral measurements was computed on the basis of the number of spectra falling within ±1 standard deviation of the MRS [5]. Spectral smoothening was carried out on MRS for noise removal using low-pass, Savitzky-Golay filters whenever required.

4. Spectral discrimination of live and bleached corals

Reflectance spectra or spectral signatures (i.e. reflected light as a function of wavelength) of live corals are considered as a fundamental parameter in reef remote sensing [22] as it is the key determinant of coral cover and coral health. In situ hyperspectral signatures are commonly analysed with respect to 'wavelength feature' approaches (i.e. spectral feature like reflectance peaks and absorption dips) where the feature is explained with the help of established knowledge on the spectral properties of the constituent materials of the target [23]. In case of a biotic, benthic substrate like corals, reflectance is a complex function of pigments, structure and morphology. The spectral characteristics of corals get determined by pigments from three different sources: (i) zooxanthellae pigments, (ii) pigments present in the ectodermal and endodermal tissues of host coral polyp and (iii) coral skeletal pigments for some species [24]. In case of in situ hyperspectral measurements, the physical distribution of pigments combined with the colony morphology of corals will affect the spectral signal received from it [24].

The live corals sampled showed a flattened response between 400 and 550 nm (**Figure 3C**) due to the contribution of strong absorption by characteristic zooxanthellae pigment called peridinin. All these live corals show triple peaks (i.e. local maxima or shoulders) at 575, 600 and 650 nm. This characteristic triple-peaked reflectance pattern was first reported in [25] and is known as 'brown mode of coral reflectance' [22]. This spectral pattern is commonly expressed by corals which visually appear in brown, red, orange, yellow or in green colours and is hence called as brown coral mode [22]. The 575 nm peak is known to be a contribution from coral-host fluorescence [22, 26] or more specifically cited as a signature of phycoerythrin (a photosynthetic accessory pigment found in red algae) fluorescence [24, 25]. *Turbinaria peltata* records this peak with a shift of 5 nm at 580 nm. Signature of phycoerythrin fluorescence at 575 nm leads to the circumstantial evidence towards the presence of this particular pigment; however, the same has yet not been concretely demonstrated for corals [24]. The second peak of 600 nm appears with a 5 nm positive shift for the sampled corals at 605 nm. The characteristic third peak occurs at 650 nm which is 'more of a shoulder than a peak' [22]. Generally, the second characteristic peak occurring at 605 nm dominates the other two peaks in terms of reflectance magnitude following the conventional trend reported globally. The intermittent depression or slight absorption feature located around 590 nm can be attributed to a shifted phycoerythrin absorption reported at 570–579 nm window [27]. All the sampled live corals showed characteristic chlorophyll absorption feature in 670–675 nm window followed by a steep red edge beyond 680 nm. These characteristic chlorophyll absorption and deep red fluorescence beyond 680 nm are attributed to endosymbiont zooxanthellar chlorophyll contribution [22, 24].

The partially bleached and bleached coral spectra in the visible region (**Figure 3C**) differ considerably from the live coral spectra in terms of their magnitude rather than the spectral shape. The partially bleached coral spectrum matches the spectral shape of the live corals with the characteristic triple-peaked pattern but steadily rises in terms of magnitude, almost double the values or more, at specific wavelengths (e.g. at 590–610 nm). Considering the topmost spectral plot as the upper limit of spectral profile of the sampled bleached coral surface, it can be commented that in terms of magnitude, the bleached coral spectrum is characteristically different than that of its live counterparts. The reflectance value of the bleached coral shoots to its maximum at 590 nm, six times as that of live coral. The bleached coral spectra rise steadily in the visible region with minor breaks of slopes located at 426, 505, 545, 556, 558, 578, 586 and 590 nm. The bleached coral spectra also show a stepped pattern of descent with breaks of slopes located at 605, 623, 628, 644, 648 and 657 nm. Thereafter it plunges down to the chlorophyll absorption trough located at 675 nm. Another prominent feature in the bleached coral spectra is the loss of the characteristic first peak, i.e. 575 nm peak, as compared to the second peak, i.e. at 600 nm. Earlier observation [10] on bleached coral spectra to be higher than that of live corals and appearing spectrally similar to bright white coralline sand holds true with these spectra too.

5. Derivative analysis

Derivative analysis is a potential tool for spectral characterization and feature discrimination in the domain of hyperspectral remote sensing. Derivatives of an original reflectance spectrum are numerically computed with respect to the wavelengths. First, second- and higher-order derivatives allow the identification of exact wavelength(s) at which the inflexion points and absorption troughs are located in the original spectrum [28]. Derivatives can resolve overlapping absorption features embedded within the zero-order spectrum. Since the late 1990s, derivatives have been applied in hyperspectral remote sensing of coral reefs. Holden and LeDrew used the first and second derivatives of in situ reflectance spectra for the identification of wavelength-specific characteristics of coral reef substrates [9]. They suggested that the first derivative spectra can be reliable means to distinguish healthy and non-healthy coral.

The first (**Figure 5**) and second derivatives (**Figure 6**) of the zero-order spectra of healthy and bleached surfaces of the sampled *Turbinaria peltata* were numerically calculated over 4 nm as finite band resolution to exaggerate the spectral shapes and enhance the subtle features [5, 19]. Derivatives are computed by dividing the difference between successive spectral values by the wavelength interval separating them [9]. This method gives the approximation of the first derivative at the midpoint of the spectral values. The rate of change in reflectance (or the slope) with respect to wavelength is represented by the first derivative spectra, while the second-order derivatives exhibit the change in slope with respect to wavelength. The

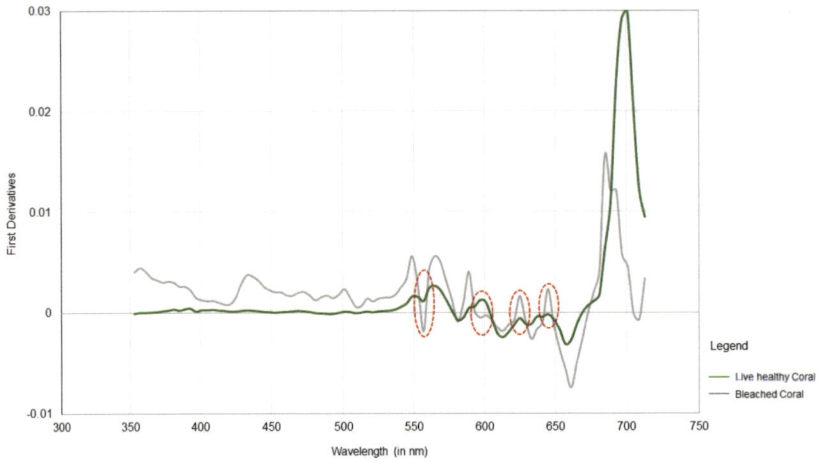

Figure 5.
First derivatives of sampled live and bleached corals (red dashed circles indicate the zones of slope differences in the derivative spectra) [5].

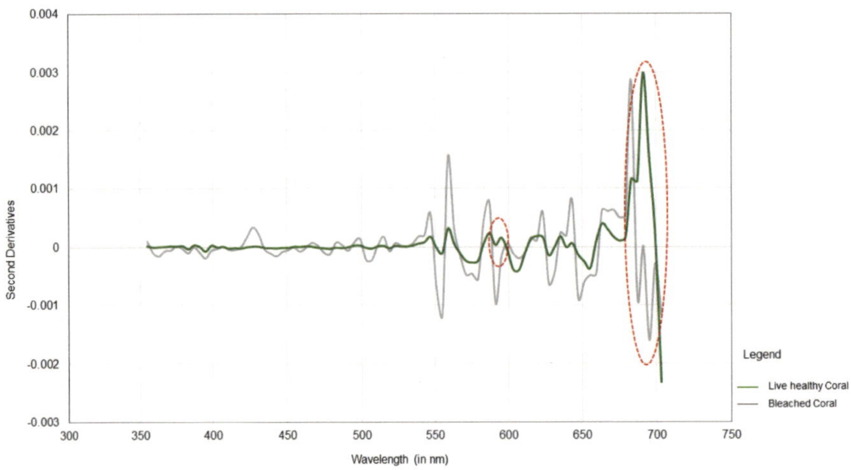

Figure 6.
Second derivatives of sampled live and bleached corals (red dashed circles indicate the zones of slope differences in the derivative spectra) [5].

first derivative spectra are considered as reliable means for spectral discrimination as they are less function of noise as compared to the second derivative spectra [5, 9, 19].

The first derivatives of sampled live and bleached coral spectra (**Figure 5**) reaffirm the magnitude difference in reflectance values in the UV-visible region (i.e. 350–550 nm). At 557 nm and at the narrow window of 593–605 nm, the first derivatives of the live and apparently healthy corals record a positive slope, while the bleached corals record negative slopes. This trend is reversed at 625 and 645 nm when the first derivatives of bleached corals record a positive slope and that of the live ones record negative slopes. The window of 593–605 nm corresponds with the observation as reported earlier [11], and it is recommended to use 596 nm (i.e. the midpoint of this window) as a slope gradient discriminator to distinguish live, recently dead and bleached corals with high accuracy. The second derivatives (**Figure 6**) of the sampled live and bleached coral spectra identify two prominent zones of slope differences, i.e. 591–599 and 687–703 nm. In both these windows, the live corals record positive second derivative values, while the bleached ones have negative values.

6. Conclusion

As observed in this study, live and bleached corals get distinguished in the visible region over 500–600 nm. The first derivatives discriminate live and bleached corals at 557, 625 and 645 nm channels and also in the spectral window of 593–605 nm. The second derivatives separate live and bleached corals in two narrow spectral windows: 591–599 and 687–703 nm.

Wavelength-specific spectral discrimination of live and bleached coral spectra using derivatives, however, needs more number of in situ data samples collected from different coral species. The onset of mass coral bleaching events can provide such ideal real-time field conditions facilitating collection of this kind of species-specific in situ reflectance data of both live and bleached corals. Field spectroscopy is a potential non-invasive tool to provide first-hand information on the health or ecological status of the corals with reference to pigment level changes at organism or colony level.

Acknowledgements

The authors are thankful to Director Shri D. K. Das, Space Applications Centre (SAC), Indian Space Research Organisation, Ahmedabad, for his encouragement and support towards coral reef targeted research programmes. The authors thank Dr. Rajkumar Sharma, Deputy Director, EPSA/SAC, and Dr. Bimal K. Bhattacharya, Division Head, Agriculture and Land-Ecosystems Division, for their constant encouragement and support to prepare this manuscript. The authors thank Dr. Sushma Panigrahy, former Group Director, ABHG/EPSA/SAC, for her guidance and inspiration to undertake this study.

The authors are thankful to the Chief Conservator of Forests, Marine National Park (MNP), Jamnagar, for necessary permission and support towards carrying out the field survey at Laku Point reef, Poshitra. The authors thank Shri Nitin Patel, Senior Research Fellow, GEER Foundation, Gandhinagar, for helping in the field data collection during the March 2011 field survey. The authors are thankful to Shri Prabakaran, C. former Research Fellow, SAC, for his kind support in field data collection. Nandini Ray Chaudhury is thankful to Dr. Dimpal Sanghvi for her kind help in the manuscript preparation.

Author details

Nandini Ray Chaudhury[1]*, Ashwin Gujrati[1], T.V.R. Murthy[1] and C.H. Satyanarayana[2]

1 Space Applications Centre, ISRO, Ahmedabad, India

2 Zoological Survey of India, Kolkata, India

*Address all correspondence to: nandinirc@sac.isro.gov.in

References

[1] Harrison PL. Sexual reproduction of scleractinian corals. In: Dubinsky Z, Stambler N, editors. Coral Reefs: An Ecosystem in Transition. Dordrecht: Springer; 2011. pp. 59-85. DOI: 10.1007/978-94-007-0114-4_6

[2] Stambler N. Zooxanthellae: The yellow symbionts inside animals. In: Dubinsky Z, Stambler N, editors. Coral Reefs: An Ecosystem in Transition. Dordrecht: Springer; 2011. pp. 87-106. DOI: 10.1007/978-94-007-0114-4_7

[3] Lesser MP. Coral bleaching: Causes and mechanisms. In: Dubinsky Z, Stambler N, editors. Coral Reefs: An Ecosystem in Transition. Dordrecht: Springer; 2011. pp. 405-419. DOI: 10.1007/978-94-007-0114-4_23

[4] Lesser MP. Experimental biology of coral reef ecosystems. Journal of Experimental Marine Biology and Ecology. 2004;**300**:217-252. DOI: 10.1016/j.jembe.2009.05.009

[5] Ray Chaudhury N. Study of some selected coral reefs of Indian coast: Based on Remote Sensing and GIS techniques [thesis]. Ahmedabad: Gujarat University; 2018

[6] Leiper IA, Siebeck UE, Marshall NJ, Phinn SR. Coral health monitoring: Linking coral colour and remote sensing techniques. Canadian Journal of Remote Sensing. 2009;**35**:276-286

[7] Reid C, Marshall J, Logan D, Kleine D. Coral Reefs and Climate Change: The Guide for Education and Awareness. Brisbane: Coral Watch, The University of Queensland; 2009. 256p

[8] Milton EJ. Principles of field spectroscopy. International Journal of Remote Sensing. 1987;**8**:1807-1827. DOI: 10.1080/01431168708954818

[9] Holden H, LeDrew E. Spectral discrimination of healthy and non-healthy corals based on cluster analysis, principal components analysis, and derivative spectroscopy. Remote Sensing of Environment. 1998;**65**:217-224. DOI: 10.1016/S0034-4257(98)00029-7

[10] Holden H, LeDrew E. Hyperspectral identification of coral reef features. International Journal of Remote Sensing. 1999;**20**:2545-2563. DOI: 10.1080/014311699211921

[11] Clark CD, Mumby PJ, Chisholm JRM, Jaubert J, Andréfouët S. Spectral discrimination of coral mortality states following a severe bleaching event. International Journal of Remote Sensing. 2000;**21**:2321-2327

[12] Yamano H, Tamura M, Kuni Y, Hidaka M. Spectral reflectance as a potential tool for detecting stressed corals. Galaxea, Journal of Coral Reef Studies. 2003;**5**:1-10

[13] Joshi D, Munjpara S, Banerji U, Parasharya D. Coral bleaching observations in the Gulf of Kachchh, India—A climate induced stress on the scleractinians. Journal of Aquatic Biology and Fisheries. 2014;**2**:106-113

[14] Patel MI. Generic diversity of Scleractinians around Poshetra point, Gulf of Kutch. Indian Journal of Marine Sciences. 1978;**7**:30-32

[15] Satyanarayana CH. Handbook on Hard Corals of Gulf of Kachchh. Kolkata: Zoological Survey of India; 2009. 114p

[16] Ray Chaudhury N, Kulkarni KM. Biokarst development: Contrasting cases from Dwarka coast, Gujarat, India. Transactions: Journal of the Institute of Indian Geographers. 2013;**35**:225-236

[17] Indian Tide Tables. Indian Tide Tables Indian and Selected Foreign

Ports. Dehradun: Survey of India; 2011. 238p

[18] Ray Chaudhury N. Spectral characteristics of selected hermatypic corals from Gulf of Kachchh, India. In: International Archives of the Photogrammetry, Remote Sensing and Spatial Information Sciences, ISPRS Congress; August 25–September 01, 2012; Melbourne, Australia: Volume XXXIX-B8, XXII; 2012. pp. 333-338. DOI: 10.5194/isprsarchives-XXXIX-B8-333-2012

[19] Chaudhury RN, Gujrati A, Murthy TVR. Water column effects on in situ spectral signatures of coral species from Gujarat coast, India. Journal of the Indian Society of Remote Sensing. 2019. DOI: 10.1007/s12524-019-00992-w

[20] FieldSpec®. FieldSpec® 3 User Manual 2006, ASD. Document 600540 Rev. F. Boulder, USA: Analytical Spectral Devices, Inc.; 2006. 100p

[21] Venkataraman K, Satyanarayana C, Alfred JRB, Wolstenholme J. Handbook on Hard Corals of India. Kolkata: Zoological Survey of India; 2003. 266p

[22] Hochberg EJ, Marlin AJ, Apprill A, Andréfouët S. Spectral reflectance of coral. Coral Reefs. 2004;**23**:84-95. DOI: 10.1007/s00338-003-0350-1

[23] Hedley JD. Hyperspectral applications. In: Goodman JA, Purkis JS, Phinn SR, editors. Coral Reef Remote Sensing. Dordrecht: Springer; 2013. pp. 79-112. DOI: 10.1007/978-90-481-9292-2_4

[24] Hedley JD, Mumby PJ. Biological and remote sensing perspectives of pigmentation in coral reef organisms. Advances in Marine Biology. 2002;**43**:277-317. DOI: 10.1016/S0065-2881(02)43006-4

[25] Myers MR, Hardy JT, Mazel CH, Dustan P. Optical spectra and pigmentation of Caribbean reef corals and macroalgae. Coral Reefs. 1999;**18**:179-186. DOI: 10.10007/s003380050177

[26] Mazel CH. Spectral measurements of fluorescence emission in Caribbean cnidarians. Marine Ecology Progress Series. 1995;**120**:185-191. DOI: 10.3354/meps120185

[27] Hedley JD, Mumby PJ, Joyce KE, Phinn SR. Spectral unmixing of coral reef benthos under ideal conditions. Coral Reefs. 2004;**23**:60-73. DOI: 10.1007/s00338-003-0354

[28] Kokaly RF, Clark RN. Spectroscopic determination of leaf biochemistry using band-depth analysis of absorption features and stepwise multiple linear regression. Remote Sensing of Environment. 1999;**67**:267-287. DOI: 10.1016/S0034-4257(98)00086-4

Chapter 8

Formation, Persistence, and Recovery of Glass Sponge Reefs: A Case Study

Jeffrey B. Marliave, Laura A. Borden, Jessica A. Schultz, Donna M. Gibbs and Glen J. Dennison

Abstract

Glass sponge reefs (bioherms) are known to occur on glacial deposits but have not previously been observed to develop on fjord bedrock ridges. It is assumed that sexual reproduction dominates reef recruitment and that sedimentation can cover intact sponge skeletons. Over a decade of scuba diving research at a small fjordic bioherm, including installation of bar-coded marker stakes, transplants of loose fragments and survey transects of substrate depth with an avalanche probe have led to new insights into the dynamics of bioherm formation and persistence. We present evidence for recovery of sponge growth from scree slopes of collapsed fragments and logged the temporal changes associated with sponge fragmentation and recovery. Bar-coded stakes were installed in 2014 to enable verification of location and sponge identity through time. Photo documentation of growth, collapse, and regrowth is presented. Research on a sponge garden on glacial sediments reveals that earliest sedimentation may center around prostrate boot sponges and bristly tunicates among the cloud and vase sponges. Although hexactinellid boot sponges do not contribute to the geologic base of bioherms, they may take part as a successional community in the substrate conditioning that could result in the genesis of a glass sponge reef or bioherm.

Keywords: bioherm, cloud sponge, reattachment, regeneration, tissue recovery, asexual reproduction, geologic reef substrate, transplants, ecological succession, temperature cycles

1. Introduction

Glass sponge reefs (bioherms) existed across present-day Europe during the Jurassic 201–145 million years ago [1], but today, only coastal British Columbia and its northern coastal boundary with Alaska have such reefs. The same glass sponges grow on rocks as sponge gardens over a broader geographic range [2]. The differences in growth and longevity of such sponges in the two types of habitat settings merit consideration. Specifically, the long-term persistence of bioherms may relate to the geomorphological base of such reefs and the growth dynamics that relate to that substrate, in contrast to the sponge garden rock substrate. The geological stabilization of glass sponge skeletons by clay sedimentation creates a nonliving base that provides stability to the living reef, stabilizes dead sponge skeletons, and enables living sponges to attach to this substrate.

Bioherms are considered "bedrock averse" in terms of underlying geologic substrate [3]; glacial sediments are the normal underlying substrate for bioherm formation. Depth data for the shallowest extent of an inshore bioherm are interpreted as indicating a bedrock substrate for at least a peripheral part of the reef we have studied. Observations at a sponge garden suggest a possible successional community including sediment-accumulating species that could develop the substrate of sedimented dictyonine sponge fragments that constitutes the geomorphologic base of a bioherm.

The present study spanned climate events that may relate to trends in sponge mortality. The El Niño of 2015/2016 started in November of 2014 and lasted until May of 2016 with anomalies as high as 2.6 (http://tinyurl.com/NOAAONI). Two modest La Niña climate events occurred from August to December 2016 (maximum anomaly of −0.7) and from October 2017 to March 2018 (maximum anomaly of −1.0).

The literature indicates that dead sponge skeletons on a bioherm become embedded intact by sediment and that new sponge growth is based on planktonic settlement of sexually produced sponge propagules that settle on dead sponge skeletons [1, 4]. Not all growth is based on newly settled sponges, since recovery of damaged sponge tissue also occurs [5]. As well, fallen fragments of cloud sponge (*Aphrocallistes vastus*) have capacity to reattach and resume growth [6]. The consequences of collapse and potential for recovery, reattachment, and continued growth have not been compared for glass sponge gardens versus glass sponge reefs. This book chapter presents a theory of scree slope drift formation at sponge reefs as a means of relatively rapid growth for bioherms and further posits that such recovery does not ordinarily occur at sponge gardens.

2. Methods

Over 100 scuba dives were conducted from 2007 to 2018 at the inshore Defence Island bioherm (**Figure 1**) in Howe Sound, British Columbia (49°34.66 N, 123°16.41 W). Difficulty in relating video frames to individually identifiable sponges led to installation of 12 bar-coded locator stakes [5] in 2014 that also had a mark to indicate the depth of the reef substrate at the time of installation. Growth and death of sponges near these marker stakes were monitored with video recordings. All photographs and videos used for analysis in this chapter include one of these locator stakes or some other landmarks such as a temperature logger, a sonde oceanographic buoy, or a pair of pipes called the "pipe vee" which had been installed in 2009 for camera positioning.

Divers with the Underwater Council of British Columbia inserted temperature loggers in the seabed of various sponge reefs in Howe Sound from 2014 through 2018, including the present site. Loggers were replaced at intervals so that most sites had nearly continuous records. The observations spanned El Niño (2015/2016) and La Niña (2016/2017, 2017/2018) climate events. Beginning in May 2014, Thermochron® iButton temperature loggers (version DS1921Z-F5; accuracy = ±1.0°C) were deployed at six sponge reefs in Howe Sound, data from four of which are presented here. Loggers were placed in among the sponges using PVC pipe holders and left there for up to 1 year, logging at intervals of either 2, 3, or 4 hours. Loggers were collected and data downloaded as often as possible, though weather and boating restricted the collection of some loggers such that gaps in data occurred. Two of the six reefs had significant gaps in data so are not presented here.

Figure 1.
Defence Island bioherm, GoPro Black (enhanced), April 27, 2018, full sun with ≤30 m visibility at all depths. The lower two video frames show the southwestern side of reef, shot from the southwest (N slope left, S slope right). Middle frame is of mid-reef, shot from southwest on the mid-line ridge of the reef. Top photo is of the northeast end of the ridge.

Data collected from 2014 to 2018 were summarized as an average temperature per date and presented graphically.

Unplanned events led to research opportunities. The pipe vee had been installed in July 2009, prior to the 2009/2010 El Niño which coincided with collapse of many sponges through 2010–2012, including formation of a drift of sponge debris that came to rest just below that pipe vee. Recovery and regrowth of that pipe vee sponge drift has since then been monitored with video recordings. Specific video frames are presented to demonstrate shape and relative size at a given date.

In October of 2016, various sponges were found cut off and lying loose after apparent contact with a sport downrigger fishing line that hit the reef. Four complete sponge bushes were fixed to the reef with PVC pipes (uppermost in **Figure 1**), and a sliced section of a sponge head was placed precisely against its intact host sponge, and the healing and reattachment were monitored with video. The transplanted sponges now provide new fixed location identifiers.

Surrounding the transplanted sponges at the shallowest ridge at the east side of the reef (**Figure 1**; top), a grid of half-meter spaced measures with an avalanche probe was conducted to measure depth of bioherm over solid rock. Previous transect measures with that probe had established that the shallowest ridge top where the transplants were installed had the least depth of reef substrate, suitable for measure with the probe. We used a 3-m avalanche probe to determine the depth of bioherm sediment accumulation on top of the rocky reef. We measured sediment depth within a 300 × 300 cm grid centered on the ridge of transplanted sponges. To ensure the measurements were 50 cm away from one another, we used a 50 × 50 cm quadrat to mark each position before inserting the probe. At each position, we inserted the probe until it hit hard substrate, recorded the bioherm depth to the nearest 5 cm, and recorded the water depth to the nearest 0.1 m. Measurements were taken on August 24, 2018, and September 14, 2018. To correct for variation in tidal height between the 2 days, we measured the water depth at one of the sampling positions twice (i.e., on both days). Because the water was 2.2 m higher on September 14, we subtracted 2.2 m from all water depth measurements taken that day. We then further corrected all measurements to zero tide by subtracting 1.1 m, which was the height of the nearest slack tide on August 24. At 5 of 42 measuring positions within our grid, we could not measure sediment or water depth without risking damage to the sponges. In those instances, we interpolated the measurements by taking the average of the nearest two measurements on either side. Data were plotted using surface plots in Microsoft Excel. The bioherm thickness and the depth of the hard substrate were plotted separately.

A sponge garden on glacial till at west Bowen Island (49°23.26 N, 123°24.76 W) was videotaped in July 2013 and August 2018, capturing images of sedimentation around glass sponges and their dead fragments, along with other marine organisms that accumulated sediment. These videos were reviewed, and all identifiable animal species in the vicinity of glass sponges were recorded. A cluster of cloud sponges (called Baker's Dozen) at this site was videotaped in 2012 and 2015 to document relative growth of various sponge bushes.

3. Results

Temperature logger data for the present study site (inshore Defence Island) and three other sites are detailed in **Figure 2**. All four sites had temperature spikes in late summer of 2015 exceeding 10.0°C, but only the inshore Defence Island site exceeded 9.0°C through spring and summer of 2015 as well, before the intensification of the 2015/2016 El Niño. Note that Passage Island is the only other bioherm site as shallow as the present study site; the location of the Passage site (49°20.27 N, 123°18.89 W) is at the south entry to Howe Sound, whereas the present Defence site is at the inner sill to the north.

At the Defence bioherm, the pipe vee scree drift formed in 2010–2012 and its subsequent recovery, growth, and loss of successive bushes was documented through 2018 (**Figure 3**). The first frame in **Figure 3** (viewed from east end) shows primarily dead sponge tissue in the center of the drift, with small tubes and mittens emerging from the dead tissue. The second frame in **Figure 3** shows the pipe vee and the drift from the east; the third is an overhead shot that includes two transplanted tubes of sponge at the downhill east and west sides (both transplants perished within 2 years). The last frame shows the central portion of the drift continuing to grow, whereas the western cluster of sponge in the foreground had fallen over in the downhill direction. That fallen cluster had grown as vertical tubes without any sideways mittens contacting the surrounding substrate. The eastern head of sponge had been hit by fishing gear in fall of 2016 and had subsequently collapsed and perished.

At the distant Passage Island bioherm site, possible debris drift formation is evident on the east slope of the reef (**Figure 4**). The lower edges of those apparent drifts in **Figure 4** have well-developed sponges, but above the middle drift in the

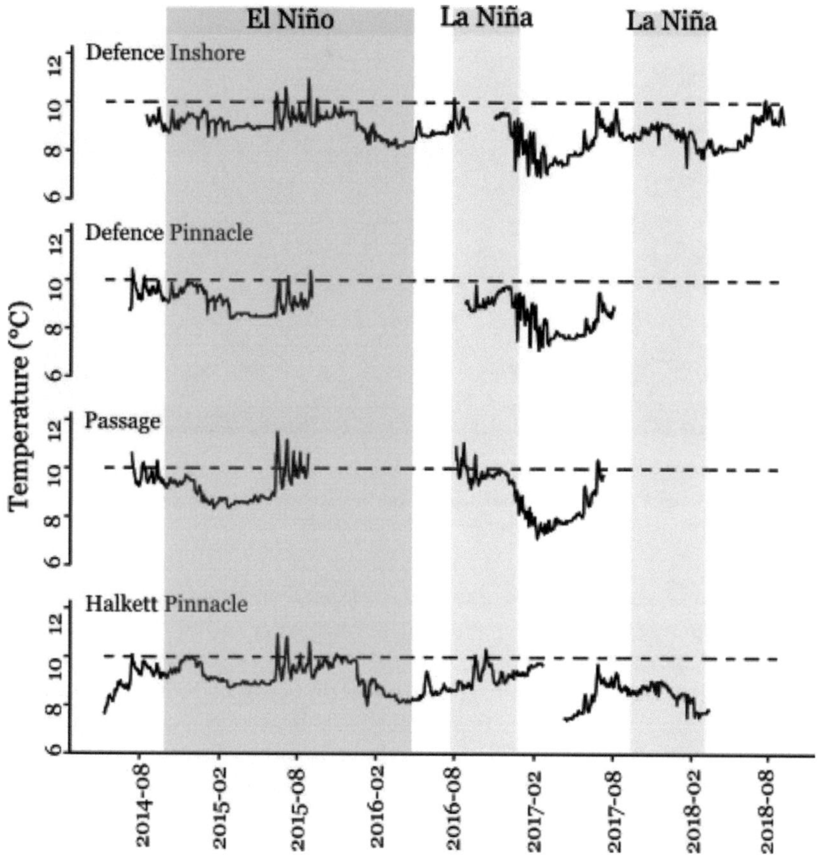

Figure 2.
Temperature at the inshore Defence Island bioherm and three similarly shallow bioherms in Howe Sound, British Columbia, May 2014–September 2018.

Figure 3.
Pipe vee sponge debris scree drift. Growth and subsequent tissue loss: top left—July 10, 2013; top right—May 1, 2014; bottom left—October 26, 2015; and bottom right—April 27, 2018. Circles identify same sponge. Note foreground center of lowest photo (arrow) shows fallen sponge from debris drift (W end).

Figure 4.
Passage Island bioherm with possible debris drifts at lower elevations on the east slope. Note debris fragments with recent growth (circled) above central sponge row at lower center.

center of the frame, there are disorganized, collapsed sponge fragments and small outgrowths of sponge that may represent tissue recovery rather than rapid growth of newly settled sponges.

The fishing gear damage to the Defence bioherm in fall of 2016 led to opportunistic transplanting of four sponge bushes at the shallowest ridge top (**Figure 5**). A loose fragment of sponge sliced from an otherwise intact host sponge bush was replaced against the host tissue and secured with stakes (**Figure 6**). By 2018 the transplants were growing and had attached to the stakes and to the underlying bioherm substrate (**Figure 5**). The staked slice of tissue was completely healed, and no sign of the slice location was evident. In September 2018, the westernmost transplant was about half dead, possibly indicating the onset of a 2018/2019 El Niño that is not confirmed on the Ocean Niño Index at the time of writing (**Figure 7**). Another

Figure 5.
Ridge top transplants (conducted October 2016), February 13, 2018. Top left photo viewed from north; top right photo viewed from south. Note attachments (circled) of sponge mittens to substrate (bottom right photo) and to stake (bottom left photo).

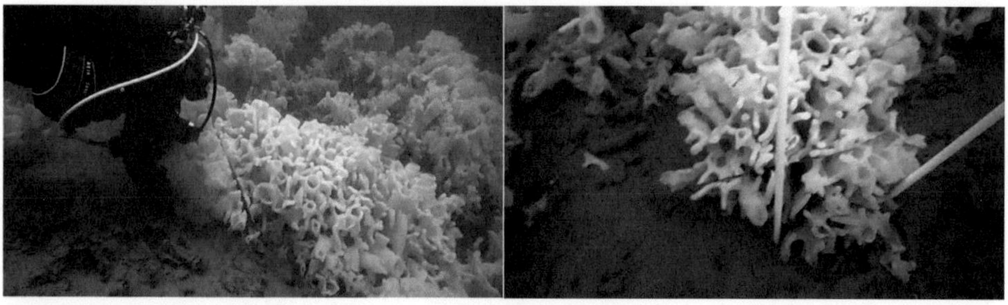

Figure 6.
Loose slice of sponge replaced against host tissue and secured with pipe stakes in October 2016.

Figure 7.
Upper photo: west transplant at ridge top dying, October 9, 2018. Lower left photo, transplant by 2-blue with dead core on July 24, 2017. Lower right, October 9, 2018, same transplant largely dead.

loose sponge head that had been staked in place at another part of the reef also died suddenly in the August/September period of 2018.

At the west Bowen Island sponge garden, four sponge heads labeled (A), (B), (C), and (D) in **Figure 8** grew relatively rapidly between September 2012 and May 2015. During fall of 2015, these sponges suffered tissue loss during the record 2015/2016 El Niño. Growth has been negligible in the subsequent 3 years.

Examples of rapid mortality at the inshore Defence bioherm were identified by means of the bar-coded locator stakes (**Figures 9–11**). At the stake with 1-black stripe (at stake top), we saw rapid necrosis of part of a sponge head over a 3-month period during spring 2015, with no subsequent spread of mortality over the next 3 years (**Figure 9**). **Figure 10** shows the sponge tissue marked at the 2-black stake in 2014 that had died adjacent to the stake by May 2015; tissue growth was evident upslope from that dead sponge by 2018. In **Figure 11** the sponge tissue at 1-green had largely died by 2017.

Upright dead sponges were observed to collapse after some period of time. At the 2-yellow stripe stake (**Figure 12**), a group of intact, dead sponges observed

Figure 8.
Baker's Dozen center sponges from north, September 26, 2012 (left); and same center sponges from north, May 4, 2015 (right). Letters identify same specific sponges.

Figure 9.
Top left—Defence 1-black: February 20, 2015; top right—May 31, 2015 (Paul Sim photo); and bottom: April 27, 2018.

during 2013 and 2014 had collapsed by 2018. The actual time required for collapse is probably very brief, as upright, dead sponges filmed on the bioherm ridge top on June 17, 2013, had collapsed by July 10, 2013.

Bioherm base layer depth at the upper ridge of the inshore Defence Island sponge reef varied from 0.30 to 1.45 m deep over an area of 3 × 3 m, with adjacent measures usually varying on the order of 0.1–0.2 m between adjacent probings at distances of 0.5 m (**Table 1**). The plot of hard bottom depth is consistent with bedrock (**Figure 13**) rather than glacial till, in which much more variation in depth would be expected on a recurring basis.

Detailed observations were made at the shallow-sloped west Bowen Island sponge garden with attention to sedimentation around dictyonine hexactinellid sponges on glacial till. Taxonomy of the community occurring at this site in 2013 and 2018 in association with the glass sponges included moderate abundances of rough patch shrimp (*Pandalus stenolepis*), galatheid crab (*Munida quadrispina*),

Figure 10.
Rapid necrosis of sponge adjacent to stake with 2-black stripes, intact February 20, 2015 (top), dead May 31, 2015 (middle—Paul Sim photo) and subsequent growth upslope (circled) from dead tissue, April 27, 2018 (note Metridium anemone and rockfish at top of stake).

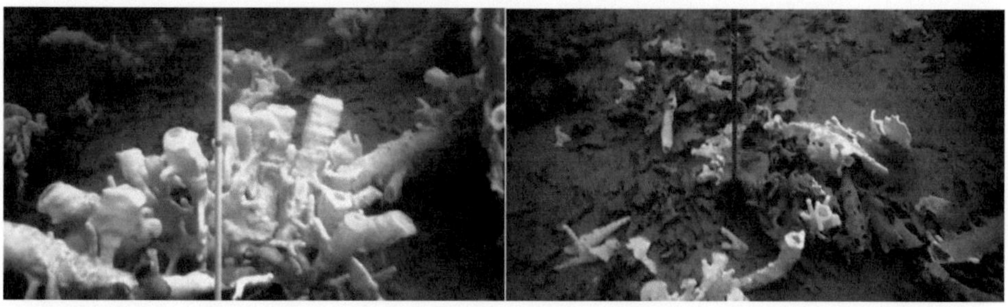

Figure 11.
Marker stake (1-green stripe): May 1, 2014 (left) and July 25, 2017 (right).

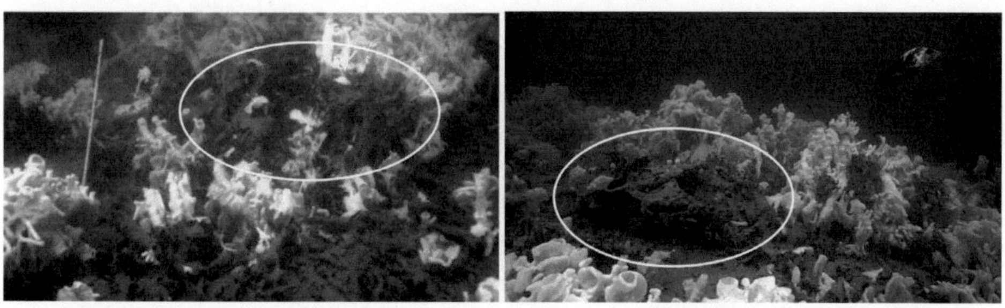

Figure 12.
Dead sponges upright at 2-yellow stake in 2014 (left) and collapsed in 2018 (right).

bristly tunicate (*Halocynthia igaboja*), fan bryozoan (*Dendrobeania murrayana*), vermilion star (*Mediaster aequalis*), white blood star (*Henricia* sp.), giant sea cucumber (*Apostichopus californicus*), and blackeye goby (*Rhinogobius nicholsii*), plus lesser abundances of 40 other animal species, not including sponges. Sponges included over 1000 cloud sponges (*Aphrocallistes vastus*) and over 1000

	A	B	C	D	E	F	G
1	110	145	140	(125)	110	115	120
2	75	70	85	(95)	105	95	95
3	70	70	60	(70)	80	90	70
4	80	70	75	(82.5)	90	70	55
5	55	65	80	(70)	60	45	55
6	80	75	65	60	55	30	40

Row and column labels correspond to the coordinates in **Figure 13**. *Numbers in brackets were interpolated based on adjacent values because live sponges growing at these locations prevented the use of the avalanche probe.*

Table 1.
Bioherm base thickness (cm) as measured with an avalanche probe.

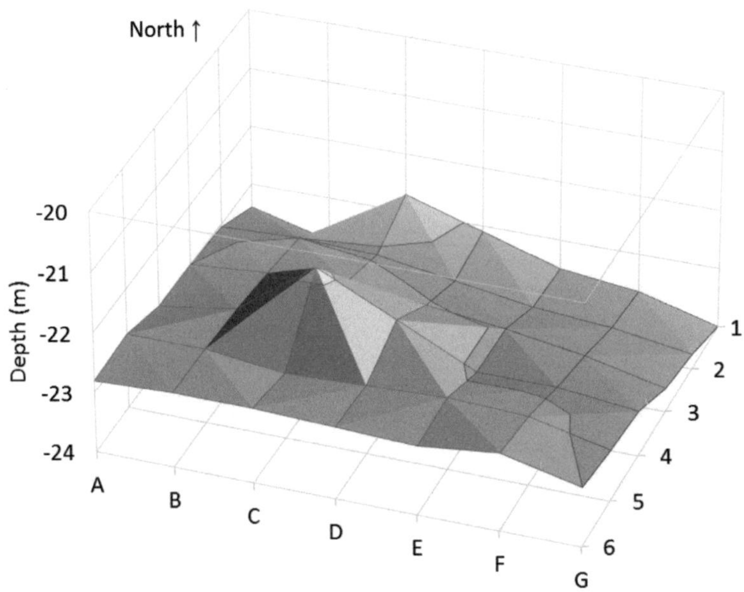

Figure 13.
Depth of the hard substrate beneath the bioherm basal layer at the Defence Island bioherm ridge. Depth was calculated by summing the water depth at zero tide and the bioherm depth, which was determined using an avalanche probe inserted into the sediment at 50 cm increments. North is toward the top and left-most position of the figure.

boot sponges (*Rhabdocalyptus dawsoni*) in each year, plus about 10 vase sponges (*Heterochone calyx*) in 2013 and over 500 in 2018, the increase in apparent numbers owing to the inability to distinguish small cloud and vase sponges in 2013. The yellow boring sponge (*Cliona californiana*) was moderately abundant, and the white meandering sponge (*Haliclona* cf. *mollis*) was at high abundance in both years. Algae included moderate abundance of red rock crust (*Hildenbrandia* spp.) and crustose corallines (*Clathromorphum*, *Lithothamnion*). Video frames in **Figure 14** lead to the hypothesis that the non-reef-forming lyssacine Hexactinellid boot sponge as well as the bristly tunicate may sequester sediment that eventually stabilizes the fallen dictyonine cloud sponge or vase sponge skeletons lying in the same vicinity. Thus, new sponges can settle on sediment-stabilized sponge fragments as well as on rock until eventually all growth might be on a bioherm-type of sedimented base composed of dictyonine sponges and their fragments.

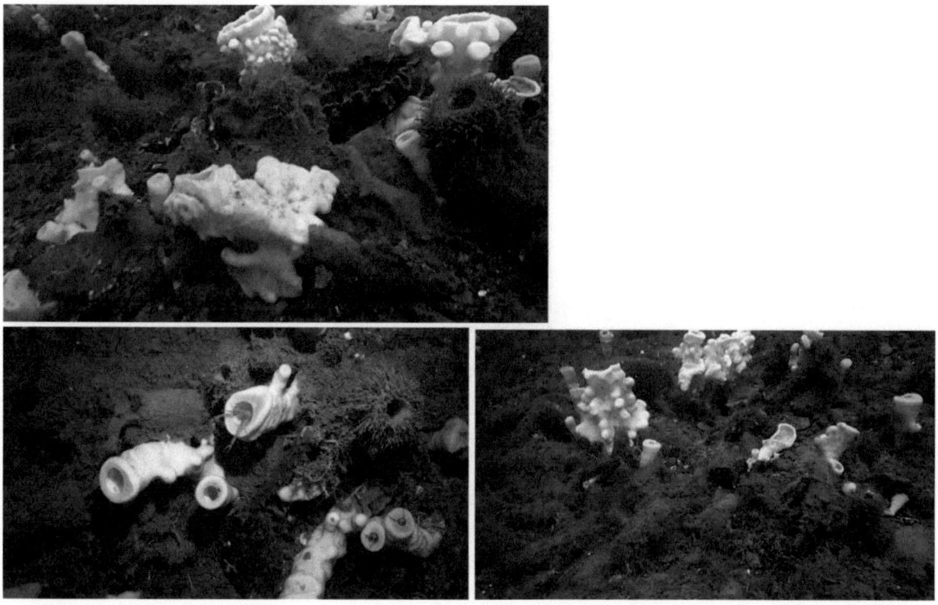

Figure 14.
Sedimentation around cloud sponges and boot sponges at the west Bowen Island sponge garden. Note sponge side attachment to both rock and dead sponge fragments in the middle photo.

4. Discussion

Glass sponge mortalities occurred at various times throughout the period of dive surveys from 2007 to 2018. Particularly extensive mortalities were associated with the El Niño climate events of 2009/2010 and 2015/2016. Note that the episode of mortality in sponge at the 1-black and 2-black stake sites occurred during the onset of the 2015/2016 El Niño during early spring of 2015 (anomalies of 0.6–1.0). This was during a period when temperatures were higher at this site than at the other three sites (**Figure 2**), whereas other sponge mortalities seemed to coincide with the heightening of that climate event during August/September 2015 (anomalies reaching 1.8 and 2.1). The coincidence of several mortality events during August/September of 2018 may enable a prediction for testing by subsequent Ocean Niño Index records for 2018/2019 if another El Niño occurs. The onset of increased mortalities in August and September of 2018 (**Figure 7**) may relate to temperature spikes associated with the onset of an El Niño climate phase at the shallow bioherms in Howe Sound. It should also be mentioned that the pH of local seawater hit low extreme levels of 7.3 in 2009 and 7.4 in 2015, compared to modal levels of about 7.7 through that period [7]; thus, low pH may interact with elevated temperature in stressing glass sponges. Interaction of climate warming and ocean acidification may soon affect shallow bioherms.

It needs to be emphasized that, without the deployment of markers [5], it would not have been possible to identify specific sponge clusters or the changes that occurred to them over time. Considerable confusion existed in trying to orient to the reef in 2012 after a period of absence during which the damage from the 2009/2010 El Niño reshaped the appearance of the reef; it was in the aftermath of that period that stakes were deployed in winter of 2013/2014. Similarly, the episode of fishing gear damage in fall of 2016 resulted in disorientation on the reef until videos of marker stakes were scrutinized and certain bushes of sponge were identified as loose and others were spotted with cuts through them. The staking of transplants in 2016 provided additional markers. Unfortunately, all these markers and oceanographic monitoring devices degrade the pristine aspect of the reef, a cost of ongoing study, but this site is remote from the more accessible bioherms used by dive charter boats in southern Howe Sound.

The depth of the present study site is 24 m at the shallowest point. The Passage Island reef is at 24 m, the Defence Island offshore reef at 31 m, and the Halkett Pinnacle reef at 32 m [8]; all other Howe Sound sponge reefs are significantly deeper (38–96 m) [8]. The temperature data (**Figure 2**) indicate that spikes in temperature exceeding 10.5°C occurred at around the heightening of the 2015/2016 El Niño in summer of 2015. By contrast, during the two subsequent La Niña years, the high temperatures were closer to 9.5°C. Cloud sponge mortalities may be associated with high temperature stress. One should note that the bioherms outside Howe Sound are uniformly deeper [1]. Therefore, the present observations represent one of two of the very shallowest known glass sponge reefs in the world, the other being Howe Sound's Passage Island reef. Similarly, the observations of a possible successional community favoring biogenic sedimentation at west Bowen Island are taken from a slightly shallower location of about 22 m depth. The much larger bioherms in northern British Columbia may have different characteristics in growth and recovery of sponge tissues. Whether a bioherm could develop at the 22 m Bowen Island sponge garden would probably depend on the future water conditions at that site. That site should be monitored for sedimentation and sponge attachment characteristics together with close attention to seawater temperature trends.

Temperatures at large sponge reefs on the continental shelf of Hecate Strait were from 5.5 to 7.3°C [9]. Those reefs occur at depths of 140–240 m, much deeper than the Howe Sound reefs at 24–96 m [8]. It is unlikely that the majority of bioherms experience such high summer temperatures as were recorded for the present study site. Therefore, the characteristics of growth and recovery of this shallowest case study site should not be predicted to establish how the deeper bioherms were formed.

If Ocean Niño climate regimes are being affected by global warming, then these two shallowest known glass sponge reefs may be predicted to be at risk of hindered persistence over time (see [10] for review). Continued study will be important to adjudicating the prospects for survival of shallower glass sponge reefs in the future. Beyond direct physiological effects of extreme temperature on glass sponges, another possibility must be considered regarding the food web of sponge reefs. The biodegradation of dying diatoms by bacteria feeds bacteria to glass sponge reefs [4]. Any temperature effect on the dynamics of diatom blooms and their subsequent biodegradation may affect the nutritional status of glass sponges, another possible impact on sponge reef health.

Sponge reefs consist of dictyonine glass sponges (hexactinellids) growing on a geologically stabilized base of glass sponge skeletal material [1, 3]. Although the lyssacine hexactinellid sponges (boot sponges) do not contribute to the sedimented geologic basal portion of bioherms, they may participate in creation of the original reef base at the level of the glacial till on which the bioherm grows. It may be the relatively flat nature of the glacial till at the west Bowen Island site, together with the turbulent drag of the cobbles and pebbles, which enables the sediment accumulation around the boot sponges and bristly tunicates at that location. Bioherms are characterized by growing on glacial till [3]. It is very likely that the inshore Defence Island bioherm is centered on glacial till because Defence Island is on the Porteau Sill, the inner sill of this fjord, the sill primarily consisting of glacially deposited boulders and cobbles. Therefore, the bedrock underneath the uppermost ridge of this bioherm probably relates to a gradual upward creep of this bioherm onto the bedrock of the island shoreline.

In this case study, glass sponge skeletons were observed to fragment significantly prior to sedimentation of the bioherm base. Thus, the increase in bioherm base depth with generations of sponges at this study site may be a minor fraction of the cumulative average height of the intact, living sponges through time. Previous literature cited here finds that deeper sponge reefs consist of frameworks of intact skeletons, and it is considered that "larger sponge skeletons may persist several hundred years before they become sedimented over" [11]. We have observed sedimentation of sponge fragments in the present study and have never observed a partially buried sponge skeleton. The portions of the sedimented bioherm base that we removed with intact living sponges attached consisted of layers of sponge skeleton lying flat. Intact skeletal specimens of siliceous sponges have been recovered from Jurassic spongiolithic deposits [12], so there undoubtedly are cases where relatively intact sponge skeletons become stabilized with sediments, but stabilization of sponge fragments must also be a component of bioherm geologic growth at the present study site. Fragmentation occurs prior to burial here, partly owing to a typically unstable tubular morphology (**Figure 15**) which may break off due to gravity after continued growth. In addition, we observe many secondary species growing on glass sponge skeletons that degrade skeletons within years (DMG unpublished data), so that collapse into fragments also relates to weakening of skeletons over relatively few years.

The assumption of sexual reproduction and reef growth from settlement of new sponges on skeletal sponge architecture [5, 11] may not be the sole source of new growth. The rapid appearance of relatively large growths out of fallen sponge fragments may have more to do with tissue recovery of stabilized fragments [6]. The experimental crushing of a portion of reef [5] did not show any recovery, but the crushing by an ROV did not leave large intact pieces of sponge as occurs in a debris drift. Many newly settled sponges on the present reef study site simply failed to thrive and disappeared. The few that survived for several years tended to obtain only modest size (<5 cm osculum diameter and height) over a period of several years (DMG, JBM personal observations). The most rapid growth we have seen has been associated with recovery or continued growth of larger sponges, including sponge fragments.

The lack of evidence for a frame-building aspect to this very shallow bioherm may indicate that it is more comparable to a biostrome [1] than to a reef mound

Figure 15.
Unstable tubular morph of cloud sponge (October 2018). This type of angled, elongate tube usually collapses by breaking off, and then dies. Note sedimentation of fallen fragments.

bioherm. It should be noted that the inshore biostromes of Queen Charlotte Sound are relatively flat [1, 13], whereas the shallowest bioherms in the Howe Sound area of the present study have much steeper top slopes than has been documented elsewhere [13]. Early work on glass sponge reefs referred to skeletal fragments in core samples [13], so the question of how intact skeletons are within all bioherms should be held moot. It remains for coring or detailed side-scan work on the shallowest inshore reefs of Howe Sound to be conducted, so whether these bioherms are any deeper in thickness than biostromes remains to be determined. Searching for evidence of frame-building within these shallow Howe Sound bioherms is another topic for future research.

5. Conclusions

A theory of scree slope drift formation at sponge reefs is proposed as a means of relatively rapid growth for bioherms; we further posit that such scree recovery does not ordinarily occur at sponge gardens owing to lack of stabilized position of fragments in garden sites resulting in continued mechanical damage of fragments in gardens. The difference is that currents continually shift and damage sponge fragments on bedrock, whereas the spicules of loose sponges can firmly lock position on a bioherm substrate under favorable circumstances of position.

The present observations indicate that a dead sponge will tend to collapse under its own weight within a few years of the death of living tissue. Similarly, growing, living sponges also have some tendency to become unstable and to collapse in the face of tidal currents, especially tubular morphs with necrosis occurring at the narrow points of attachment to the reef base. Thus, the view that bioherm growth consists of sedimentation of intact, dead skeletons of glass sponges does not fit well with our case study in which episodes occurred of necrosis and collapse of significant portions of the reef. Both the 2009/2010 and 2015/2016 El Niños coincided with certain areas of the reef dying and collapsing. The tissue collapse episode after 2009/2010 led to at least one debris drift forming and subsequently recovering growth (**Figure 3**). This is consistent with results from experimental transplant of fragments [6].

As well, a theory of successional community contribution to bioherm formation is based on observations of an extant ecological community on glacial till where sediment sequestering species of lyssacine sponge and tunicate buildup sediments in which fragments of dead dictyonine glass sponges rest secure from movement by seabed tidal currents. Attachment of live dictyonine sponges to such stabilized dead fragments can occur (**Figure 14**). A major question concerns what variation in rates of sedimentation can occur during formation of sponge reefs. A related question concerns what the dynamic is between ongoing vertical growth, collapse, and regrowth of the living surface of the reef in relation to the vertical growth of the geologically stabilized, sedimented reef base. The present study suggests that the living reef has a more dynamic range of growth, collapse, and regrowth through time than has been presumed.

These very shallow sponge reefs in Howe Sound, the only cases in the world amenable to studies based on scuba diving with compressed air, may afford valuable opportunities for citizen science contributions based on video recordings. It must be cautioned, however, that without landmarking against either geologic features or use of marker stakes, it is nearly impossible to prove identity of a sponge from one point in time to another owing to significant changing of shape and size.

Acknowledgements

Volunteer divers assisted the efforts of the Underwater Council of British Columbia (GJD) with temperature logger installation and recovery at various reefs. Volunteers also assisted in work at the inshore Defence Island reef studied by the Coastal Ocean Research Institute (CORI) team (JBM, LAB, JAS, DMG). Amanda Weltman, Boaz Hung, Jeremy Heywood, Conor McCracken, Danny Kent, and Jonathan Wong assisted the CORI authors with various aspects of the work. We thank Anya Dunham and Kim Conway for critical reviews. Partial support of work by the CORI team included grants from Mountain Equipment Co-op and the Environmental Damages Fund (Environment Canada).

Conflict of interest

The authors have no conflicts of interest.

Author details

Jeffrey B. Marliave[1*], Laura A. Borden[1], Jessica A. Schultz[1], Donna M. Gibbs[1] and Glen J. Dennison[2]

1 Ocean Wise Coastal Ocean Research Institute, Canada

2 Underwater Council of British Columbia, Canada

*Address all correspondence to: jeff.marliave@ocean.org

References

[1] Krautter M, Conway KW, Barrie JV, Neuweiler M. Discovery of a “living dinosaur”: Globally unique modern hexactinellid sponge reefs off British Columbia, Canada. Facies. 2001;**44**:265-282

[2] Marliave JB, Conway KW, Gibbs DM, Lamb A, Gibbs C. Biodiversity and rockfish recruitment in sponge gardens and bioherms of southern British Columbia, Canada. Marine Biology. 2009;**156**:2247-2254

[3] Shaw J, Conway KW, Wu Y, Kung R. Distribution of Hexactinellid Sponge Reefs in the Chatham Sound Region. British Columbia: Geological Survey of Canada. Current Research 2018-1. 14 p; 2018. DOI: 10.4095/306310

[4] Kahn AS, Yahel G, Chu JWF, Tunnicliffe V, Leys SP. Benthic grazing and carbon sequestration by deep-water glass sponge reefs. Limnology and Oceanography. 2015;**60**:78-88. DOI: 10.1002/lno.10002

[5] Kahn AS, Vehring LJ, Brown RR, Leys SP. Dynamic change, recruitment and resilience in reef-forming glass sponges. Journal of the Marine Biological Association of the UK. 2016;**96**:429-436. DOI: 10.1017/S0025315415000466

[6] Marliave J. Cloud sponge, *Aphrocallistes vastus* (Porifera: *Hexactinellida*), fragment healing and reattachment. The Canadian Field-Naturalist. 2015;**129**(4):399-402

[7] Marliave J, Gibbs DM, Young S, Borden L. 2011 climate regime: Revealed by seabed biodiversity. In: Chandler PC, King SA, Boldt J, editors. State of the Physical, Biological and Selected Fishery Resources of Pacific Canadian Marine Ecosystems in 2017. Canadian Technical Report of Fisheries and Aquatic Sciences 3266. Viii. 2018. 245 p

[8] Clayton L, Dennison G. Inexpensive video drop-camera for surveying sensitive benthic habitats: Applications from glass sponge (*Hexactinellida*) reefs in Howe sound, British Columbia. The Canadian Field-Naturalist. 2017;**131**(1):46-54. DOI: 10.22621/cfn.v131i1.1783

[9] Whitney F, Conway K, Thomson R, Barrie V, Krautter M, Mungov G. Oceanographic habitat of sponge reefs on the western Canadian continental shelf. Continental Shelf Research. 2005;**25**:211-226

[10] Conway KW, Whitney F, Leys SP, Barrie JV, Krautter M. Sponge reefs of the British Columbia, Canada coast: Impacts of climate change and ocean acidification. In: Carballo JL, Bell JJ, editors. Climate Change, Ocean Acidification and Sponges. Berlin: Springer; 2017. pp. 429-445. DOI: 10.1007.978-3-319-59008-0_10

[11] Krautter M, Conway KW, Barrie JV. Recent hexactinosidan sponge reefs (silicate mounds) off British Columbia, Canada: Frame-building processes. Journal of Paleontology. 2006;**80**:38-48. DOI: 10.1666/0022-3360(2006)080[0038:RHSRSM]2.0.CO;2

[12] Leinfelder RR, Krautter M, Laternser R, Nose M, Schmid DU, Schweigert G, et al. The origin of Jurassic reefs: Current research developments and results. Facies. 1994;**31**:1-56

[13] Conway KW, Barrie JV, Austin WC, Luternauer JL. Holocene sponge bioherms on the western Canadian continental shelf. Continental Shelf Research. 1991;**11**:771-790

Chapter 9

Artisanal Harvest of Shellfish in the Northeastern Atlantic: The Example of Limpet and Topshell Fisheries in the Archipelago of Madeira

Ricardo Sousa, Rodrigo Riera, Joana Vasconcelos, Lídia Gouveia, Ana Rita Pinto, João Delgado, Adriana Alves, José A. González, Mafalda Freitas and Paulo Henriques

Abstract

The harvesting of littoral benthic shellfish in the archipelago of Madeira dates back to the fifteenth century when the Portuguese discovered and colonized the archipelago. The consumption of littoral shellfish is part of the gastronomic cultural heritage of this region, appreciated by the local population and tourists, and has a high social and economic importance. Therefore, harvesting pressure on these resources is one of the greatest concerns, and as such, a sustainable exploitation based on proper regulation, considering the biological and ecological specificities of these species in their particular habitat, is crucial to promote the preservation of species and habitats at medium and long terms. This study presents the current harvesting management regime for gastropods in the archipelago of Madeira and characterizes the artisanal harvest through a period of 27 years (1990–2017) providing new insights for future research in these topics. This artisanal harvesting operates mostly by small vessels (<10 m), with low tonnage and capacity, in nearby areas preferentially in the North coast of Madeira and around Desertas Islands. During the studied period, management actions resulted in the reduction of 50% of the vessels operating in the harvesting of limpets and in slight recovery of the stocks of limpets. The economic impact of limpets gradually increased over the years, representing in 2017 96% of the economic value landed for molluscs and 2% of the total landings in this region. The present characterization provides a comprehensive outlook of the evolution of the marine gastropod harvest in the archipelago of Madeira and allows future comparisons with other regions where gastropods are commercially exploited.

Keywords: harvesting, management, sustainable exploitation, limpets, topshells, archipelago of Madeira

1. Introduction

The artisanal fisheries of marine invertebrates in the archipelago of Madeira (NE Atlantic) target mainly gastropods and cephalopods. The main gastropod species harvested are the limpets *Patella aspera* Röding, 1798, and *Patella candei* d'Orbigny, 1840, the topshell *Phorcus sauciatus*, and the whelk or redmouthed rocksnail *Stramonita haemastoma*. Regarding cephalopods, the common octopus *Octopus vulgaris* Cuvier, 1797; the squid European *Loligo vulgaris* Lamarck, 1798; and the orangeback squid *Sthenoteuthis pteropus* (Steenstrup, 1855) are the main target species of this fishery. Additionally, the cuttlefish *Sepia officinalis* Linnaeus, 175, and the European flying squid *Todarodes sagittatus* (Lamarck, 1798) are sporadically captured.

The artisanal harvest of gastropods in the archipelago of Madeira is a low-cost activity, usually carried out by the owner of the vessel accompanied by professional snorkelers. This activity is one of the most important small-scale fisheries in this region, due to the economic and social benefits it provides directly to the coastal communities and indirectly to the whole community. This fishery dates back to the fifteenth century when the Portuguese colonized the archipelago. The good accessibility to the rocky shores prompted the exploitation of marine shellfish resources along the coast. The harvest activity becomes progressively more intensive with the demographic increase of human settlement around the islands' coasts and with the technological progresses that simplify the access to the coast at previously inaccessible areas [1, 2]. This long-term exploitation has changed the population dynamics, due to shifts on the abundance and/or size structure and density of the exploited marine gastropods over the years [3].

In the past, four species of the genus *Patella* were reported for the archipelago of Madeira, namely *Patella aspera* as the most abundant species, *Patella caerulea*, *Patella lusitanica*, and *Patella vulgata* (known locally as "concharéu" due to its large size, sharp edges, and helmet shape). The distribution of *P. candei* (formerly identified as *P. vulgata*), which once occurred in all the islands of the archipelago, became restricted, in the beginning of the twentieth century, to the Selvagens Islands [1]. Presently, *P. aspera* and *P. piperata* (formerly identified as *P. lusitanica*) are common species in all islands from the archipelago of Madeira, *P. candei* (formerly identified as *P. caerulea*) is present in Madeira Island, Porto Santo, and Desertas, while the real *P. candei* is restricted to the Selvagens Islands. Recently new molecular tools developed with novel microsatellite markers using next-generation sequencing suggest the use of *P. candei* for the species from the Selvagens and *Patella ordinaria* for the species of Madeira, Desertas, and Porto Santo islands [4].

Concerning topshell exploitation, two species of the genus *Phorcus* have been harvested in Madeira archipelago since early colonization times. *Phorcus sauciatus* (formerly identified as *Trochus colubrinus* Gould, 1849), the most common species and with a wider geographical distribution, occurring in all islands of the Madeira archipelago, and *Phorcus atratus* are restricted to the Selvagens Islands as the endemic subspecies *Phorcus atratus selvagensis* [5]. Nowadays, *P. sauciatus* continues to be exploited in Madeira, Porto Santo, and Desertas, except in the marine protected areas (MPAs), and *P. atratus selvagensis* is not commercially exploited since its distribution is restricted to the MPA of the Selvagens where harvesting is not allowed.

Both limpet consumption and topshell consumption in Madeira archipelago are part of the gastronomic cultural heritage of this region, appreciated by the local population and tourists alike, and have a high socioeconomic importance. Therefore, harvesting pressure on these resources is one of the greatest concerns. As such, a sustainable exploitation, based on suitable regulation considering the biological and ecological specificities of these species in their particular habitat, is crucial to promote the preservation of species and habitats at medium and long terms.

The aim of present work is to compile and characterize the harvest of limpets and topshells in the archipelago of Madeira, considering and discussing the evolution of the landings and economic values and describing the activity, fishing fleets, exploited species, and yields for a period of 27 years (1990–2017). Additionally, the impact of this activity on selected aspects of limpet and topshell population dynamics is analyzed and discussed.

The implemented management measures regulating the harvest of marine molluscs in the archipelago of Madeira are characterized in detail and their impact on the exploited stocks critically discussed through a comparative analysis of any relevant available data on these species from this region. Finally, the economic and social contextualization of this fishery is made in the overall fisheries sector in the region.

2. Study area

The archipelago of Madeira is located in the northeastern Atlantic Ocean and is included in the Macaronesian biogeographical region together with the Azores, Canary, and Cape Verde (**Figure 1**). The islands of these archipelagos are of volcanic origin, resulting from the activity of several geological hotspots and sharing the oceanic nature, the geographic location, and the climatic regime. However, with specific characteristics according to the proximity of the islands to the mainland regions [6].

The archipelago of Madeira comprises the islands of Madeira (741 km^2 of area), Porto Santo (42 km^2), Desertas (14 km^2), and the Selvagens islands (3 km^2). The island of Madeira is located approximately at 635 km from Morocco and at 900 km from the Portuguese mainland. The Selvagens islands are the southernmost territory of Portugal at 239 km from the island of Madeira and at 375 km from the coast of Morocco. This subtropical archipelago is influenced by the Azores anticyclone, the Gulf and the Canary currents, the continental anticyclonic center of Northwest Africa and Western Europe, and the frontal systems associated with the lower pressure center of the polar front [7].

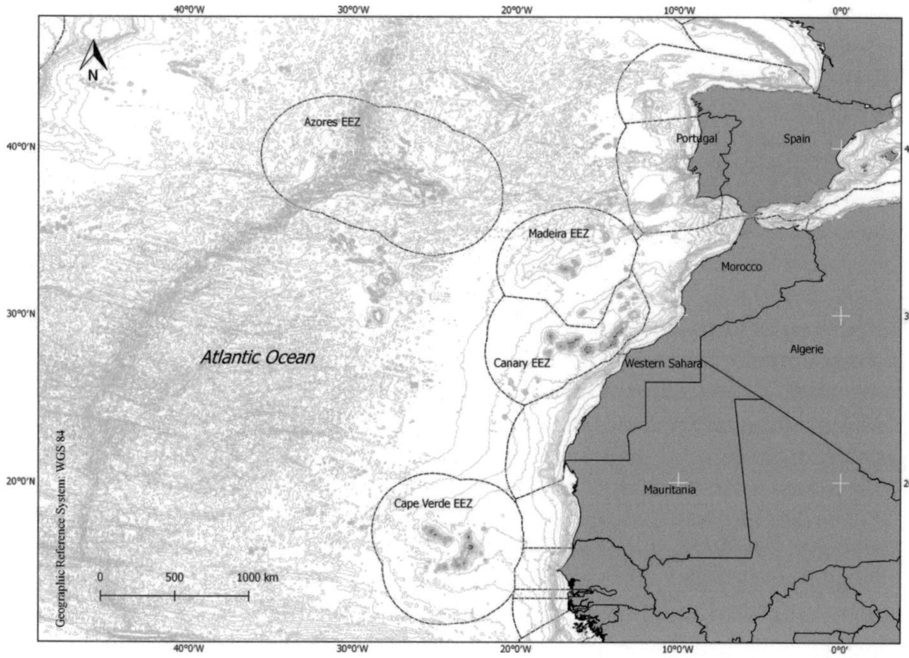

Figure 1.
Representation of the southern part of the Northeastern Atlantic showing the study area, the archipelago of Madeira, included in the Macaronesian biogeographical region.

The islands of Madeira and Desertas represent the most recent islands of the archipelago of Madeira with 4.6 and 3.6 million years, respectively. Porto Santo has an estimated age of 14.3 million years and the Selvages islands an estimated age between 24 and 29 million years [8, 9].

The population living in the archipelago of Madeira in 2017 was *ca.* 254 thousand inhabitants, and the fishing activity employed 618 registered fishermen (DRP-RAM). The annual landings increased 38.6% in relation to 2016 with 6.739 tonnes of fish and molluscs corresponding to 21,636 thousand €.

2.1 Data collection

Data on the landings (i.e., species, day, weight, and economic value) and on the artisanal fleet (i.e., length of the fishing vessel, tonnage, capacity, métier, fishery license) were obtained from the Regional Fisheries Department of Autonomous Region of Madeira (DRP-RAM) for both limpets and topshells.

Logbook data analyses were only available to characterize the limpet harvesting activity (i.e., harvesting area, typology of bottom, depth of harvesting, number of divers, number of snorkelers per vessel, and landing place and time), since it is mandatory to fill the logbooks according to the regulation of the limpet harvest in the archipelago of Madeira. For topshells these data are not available due to this activity not being regulated.

Landings data were compiled and analyzed covering the period from 1990 to 2017 and the logbook data from 2008 to 2017, concerning the period after the regulation of the harvesting of limpets that required the filling of all the harvesting information in logbooks.

3. Harvesting regulation in the archipelago of Madeira: a driver to sustainable exploitation

The fisheries sector should protect fish resources and environment through an effective legal regulation and an appropriate compliance and enforcement to ensure the sustainable resource exploitation [10]. Based on this principle, the regulation of limpet harvesting in the archipelago of Madeira sets the basis for a sustainable and responsible exploitation of these resources.

The rules of governing the harvesting of limpets in Madeira were initially set by the necessity to establish proper measures to the regional specificities, with regard to the practice of underwater hunting. These measures were implemented through the article 6 of the Regional Government under the Regional Legislative Decree No 11/1995/M on 21 June 1995 and setting a maximum allowable catch of 3 kg/day per person for limpets. This was the first management action regarding the harvesting of gastropods in the archipelago of Madeira.

The legal regulation of the limpets' harvesting method was decreed by the Legislative order No 1102-B/2000, 22 November 2000, which established the tools and instruments to be used in limpet harvesting.

The current limpets' management in Madeira was based in technical measures implemented by the Regional Government under the Regional Legislative Decree No 11/2006/M, 18 April 2006, which establishes the legal regime for the harvesting of these gastropods in this region [3]. The management measures were implemented in 2006 based on the knowledge obtained from studies on the biology, population dynamics, and assessment of the stocks and intended to advise a precautionary approach capable of harmonizing the need to protect stocks with the preservation of the economic activities associated with their capture and gastronomic use [11].

For this purpose, regulators established several management measures for the traditional and commercial exploitation of the species *P. aspera* and *P. candei*. For the commercial exploitation, a minimum catch size of 40 mm, the obligation of harvesting licenses, logbook provision, the landings and first auction sale, and catch limits were implemented, enforcing the maximum allowable commercial catch of 15 kg/person/day or 200 kg/boat/day, being exempted of any license the traditional harvest that does not exceed 3 kg/day per person.

Additionally, the Legislative Orders No 80/2006 and No 81/2006, 17 June 2006, set the establishment of a closed season between November 1 and January 31 to avoid limpet harvest during the reproductive season and the rules concerning the harvesting card.

In 2009, based on continuous population monitoring, the closed season was changed to become effective between December 1 and February 28 (Legislative Order No 5/2009, 4 July 2009). In 2016, the closed season was extended through the Legislative Order No 40/2016, 17 February 2016, in result of the data obtained from the continuous monitoring of the stocks, in order to more efficiently provide protection to these heavily exploited species, now lasting from December 1 to March 31. The extension of the closed season is intended to result in biological benefits, allowing greater protection of spawning, larvae development, and settlement and to increase the success of annual recruitment and subsequent development of limpets. Additionally, the competent authorities implemented a reduction in the number of harvesting licenses aiming to diminishing the pressure on the resource [3, 12].

Since the implementation of limpet harvest regulation, the non-compliance of the imposed management measures is punishable with administrative offenses and penalties. The sanctions include monetary fines (between 49.88 € and 44,891.81 €), harvesting prohibition, and suspension of harvesting licenses. Nevertheless, poaching continues to occur during the closed season, without abiding the minimum catch size of 40 mm of shell length [3].

The regulation of limpets in the archipelago of Madeira allowed safeguarding the peak of reproduction, the immature specimens, and the catches of both *P. aspera* and *P. candei*. Although the management measures on these resources have been implemented for several years, only a slight recovery on the limpet populations has been observed, maybe due the peculiar life trait characteristics of this species, as slow growth and longevity. Hence, it is of paramount importance to keep up with the enforcement of the implemented measures in order to achieve higher recovery rates of the exploited limpet stocks.

Currently, topshell harvest in the archipelago of Madeira is not regulated, with the exception of harvest ban on MPAs [2]. As such, the effort exerted and the shifts on the populations' dynamics of the harvested populations, exploited for more than 500 years, are unknown. However, management measures based on recent studies [2] are in progress and shall enter into force during the year 2019, aiming to promote both profitable and sustainable harvest. The establishment of a maximum catch of 2 kg per day for noncommercial use and 20 kg per day for commercial use, the implementation of landing obligations and first sale at auction, the establishment of a minimum catch size of 15 mm shell length, and the establishment of a closed season between February and May are recommended [2].

Further European legislation through the European Community Council Regulation No 199/2008, 25 February 2008, prompted an increase in the knowledge regarding the fisheries sector in the European Union, including smaller sections such as limpet harvest, meeting the demands generated by the necessity to evolve toward a sustainable fisheries sector, with its management based on the fleet and fishing areas rather than based on fish stocks. As such, the collection of data and their availability by region will provide the basis for the better scientific advice.

4. Harvesting

Limpets and topshells are collected by hand, in the intertidal zone by the local population (traditional harvesting) and in the subtidal zone by snorkelers executing several dives per day (commercial harvesting), from 1 to 6 m deep ($\bar{x}$ = 1.74 ± 1.31). The harvesting fleet operates preferentially on the northern coast of Madeira and Desertas Islands (**Figure 2**), the least accessible zones, while traditional harvesters collect limpets from all around the island preferably in areas with easy access and milder sea conditions.

The harvest of limpets in the archipelago of Madeira is operated from April to November for limpets and all year round for topshells mostly by small vessels (<10 m). Between 1990 and 2017, the number of vessels operating on the harvesting of gastropods was reduced from 17 (1990) to 9 (2017) (**Figure 3**). The observed reduction in the number of vessels and licenses results from the implementation of regulation, which among other management measures included the gradual reduction in the number of fishing licenses and vessels in order to reduce the harvesting pressure on limpet stocks.

The majority of the fishing vessels have less than 10 m length (78%), and the remaining ranges between 10 and 12 m (11%) and between 12 and 18 m (11%). This artisanal fleet includes vessels of low tonnage (0.74–17.28 gross tonnage) and capacity between 11 and 136 KW, usually operating in nearby areas reachable in a short time.

Limpet harvest is deeply rooted in the local community and particularly on coastal fishing communities representing an additional revenue source for many families, thus contributing to the local economy. The number of snorkelers per fishing vessel, between 2008 and 2017, varied between two and eight depending on the vessel size ($\bar{x}$ =4.9 ± 1.29). The number of professional harvesters registered per vessel and nonprofessional harvesters (15 kg/person/day) decreased over the study period following the management measures implemented (**Figure 4**).

A remarkable proportion of the Madeiran fishing fleet focuses solely on the harvesting of gastropods; nevertheless due to the implementation of a closed season, some fishing fleets operate with two or three métiers. In 2017, 56% of the fishing fleet operated exclusively in the harvesting of gastropods; 33% operated with 3 métiers and 11% with 2 métiers (**Figure 5**). The complementary métiers were essentially directed to the capture of tuna fish and demersal fish.

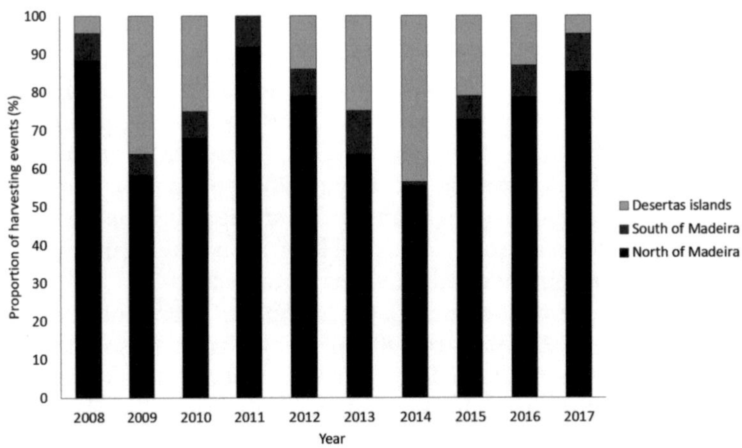

Figure 2.
Proportion of harvesting events of limpets per zone based on data from logbooks.

Figure 3.
Number of fishing vessels operating in the harvesting of limpets from 1990 to 2017 in the archipelago of Madeira.

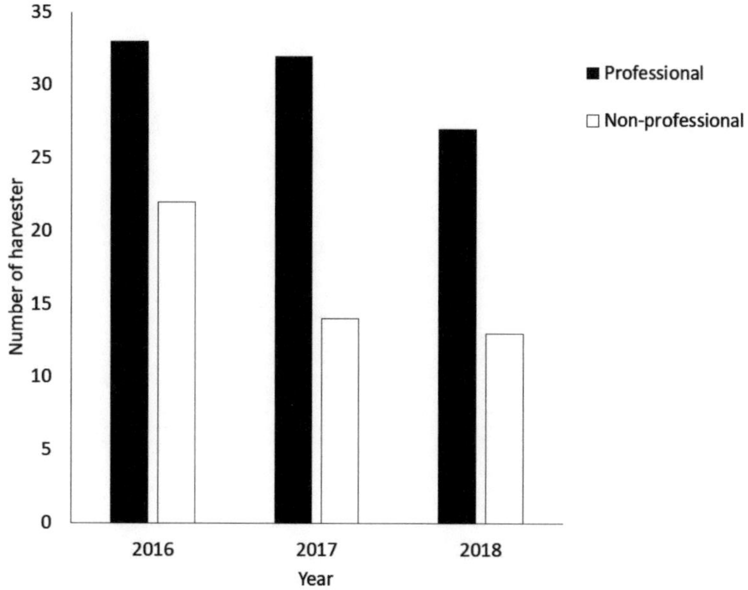

Figure 4.
Number of professional and nonprofessional harvesters registered in the archipelago of Madeira between 2016 and 2018.

4.1 Species and yields

In the archipelago of Madeira, limpets are mostly harvested by scuba diving in a mixed exploitation of *P. aspera* and *P. candei* with relevant commercial importance [13, 14] for the involved population. This activity represented approximately 1.5% of the total of the fisheries and 2% of the total of the economic value (€) in 2017, reaching an average value of 4 € per Kg [3]. In 2011, this activity represented *ca.* 5% of the total economic value (€) of the fisheries which landed in the archipelago of Madeira.

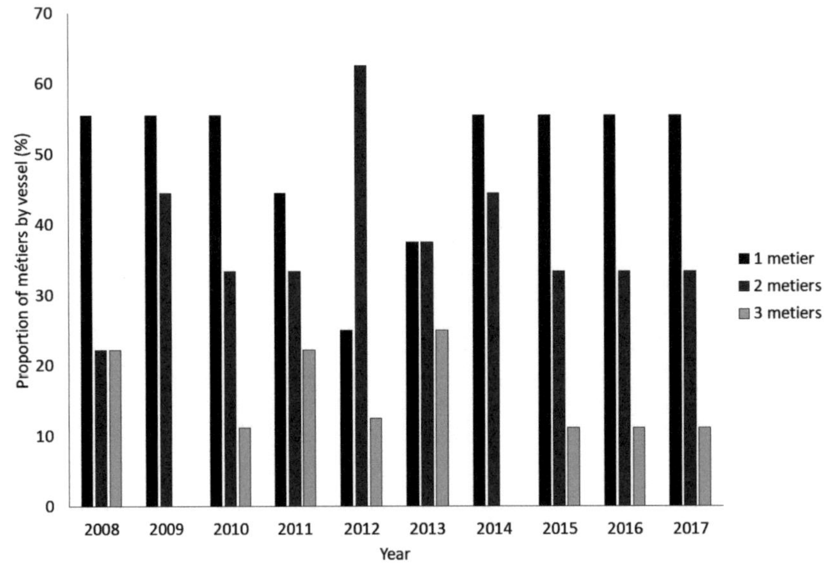

Figure 5.
Proportion of métiers operated by the fishing vessels involved in the harvesting of gastropods in the archipelago of Madeira.

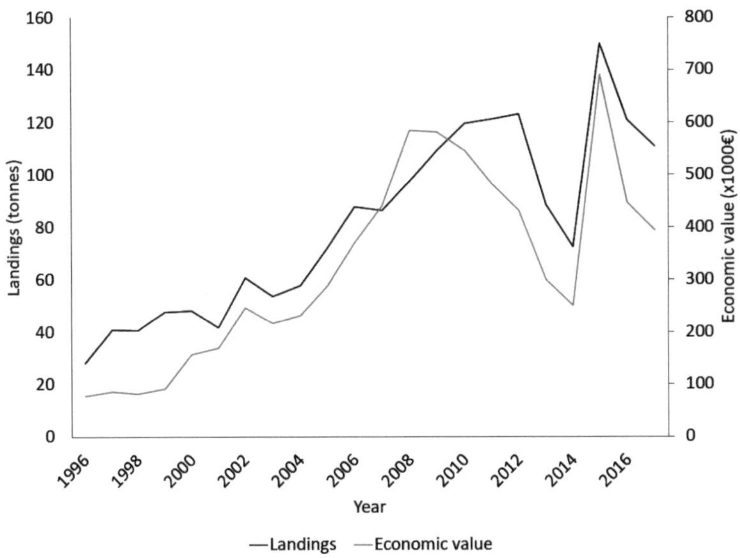

Figure 6.
Representation of landings (tonnes) and economic value (1000 euros) of limpets from 1990 to 2017 in the archipelago of Madeira.

The commercial landings in weight of limpets harvest varied from *ca.* 5 tonnes in 1990 to *ca.* 111 tonnes in 2017 (**Figure 6**). The maximum value landed was 150 tonnes in 2015 yielding a first auction sale value of 0.7 M€. The decrease in landings in 2016 and 2017 is related to the increase of the closed season from 3 to 4 months, reducing the harvesting activity in 1 month which was reflected in the annual landing values. Data from 1990 to 2006 needs to be analyzed carefully since limpet harvesting was not regulated and landing obligation was not mandatory.

The limpets landed in Madeira are for internal consumption and, to a minor extent, for exportation, mainly to the archipelago of Azores. However, data on the exportations of limpets are not available and as such are not possible to determine accurately.

The average annual value of limpets per kg landed in the archipelago of Madeira fluctuated considerably, from 1990 to 2017, increasing from 1.80 € in 1990 to 3.65 € in 2017 ($\bar{x}$ =3.33 ± 1.25 €). For this period the average annual price varied from 1.21 € in 1990 to 5.90 € in 2008 (**Figure 7**). Since 2008, the value per kg decreased due to the increase in landings and the demand for this resource. Even so, the value per kg is still very attractive for the professionals involved in this activity.

The harvesting effort, represented by the number of days employed in the harvest of limpets, decreased from 1448 days in 2008 to 655 days in 2017. However, in general the reduction in fishing effort contrary to what would be expected led to an increase in the landings (*ca.* 98 tonnes in 2008 to *ca.* 111 tonnes in 2017). From 2013 to 2017, the oscillations on the landings were related to different sea conditions among years, e.g., in 2015 the milder sea conditions resulted in more 361 days at the sea. The increase verified in the landings seems to result from the management measures implemented in 2006 (**Figure 8**). In fact, if the management measures implemented were successful, then it is expected that the exploited stocks have greater biomass, not only in density but also larger individuals are supposedly more abundant which in return means that lower effort returns similar or superior yields.

The highest proportion of landings of limpets, for the considered time period, occurred from May to August, corresponding to the months with better sea conditions in the archipelago of Madeira (**Figure 9**). The comparison between monthly landings before and after the implementation of the closed season showed that the highest proportion of landings remained identical between the two periods, except for the months of the closed season, when limpet harvesting is not allowed.

In the archipelago of Madeira, landings of limpet commercially harvested, between 1990 and 2017, occurred in 10 ports, namely, Funchal, Câmara de Lobos, Calheta, Madalena do Mar, Paúl do Mar, Porto Moniz, Santa Cruz, Machico, Caniçal, and Porto Santo. The total landings per port in tonnes are represented in **Figure 10**. During this period, Porto Moniz, Funchal, Caniçal, and Paúl do Mar were the most important ports representing approximately 96% of the limpets landed in 1990 and 100% of the limpets landed in 2017.

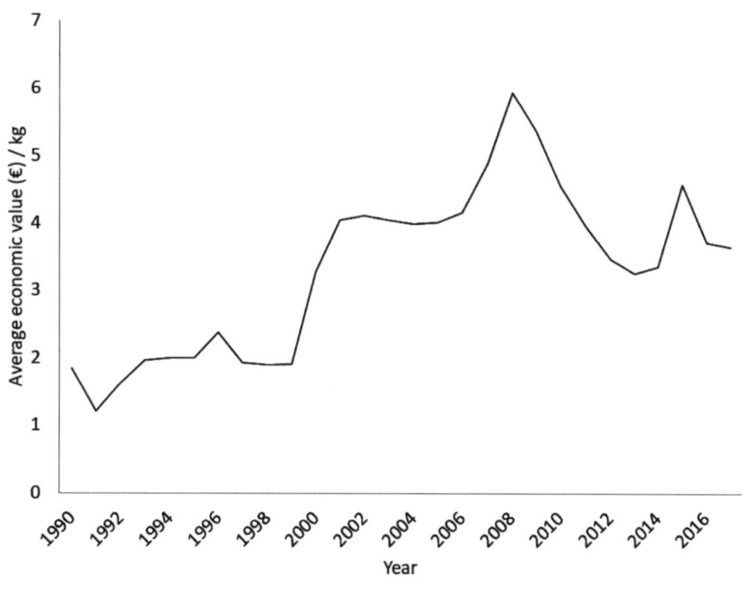

Figure 7.
Average annual price of the limpets landed in the archipelago of Madeira from 1990 to 2017.

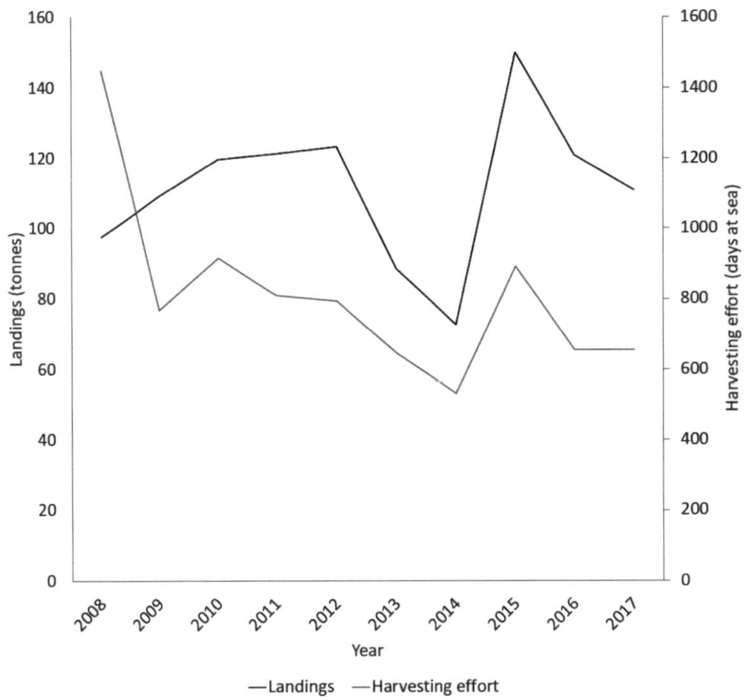

Figure 8.
Landings and harvesting effort from 2008 to 2017 in the harvesting of limpets in the archipelago of Madeira.

Porto Moniz showed the highest annual limpet landings with 1.9 tonnes in 1990 and 76 tonnes in 2017, representing approximately 69% of the total landings of limpets in 2017 (**Figure 11**). Six fishing vessels landed *ca.* 96 tonnes representing

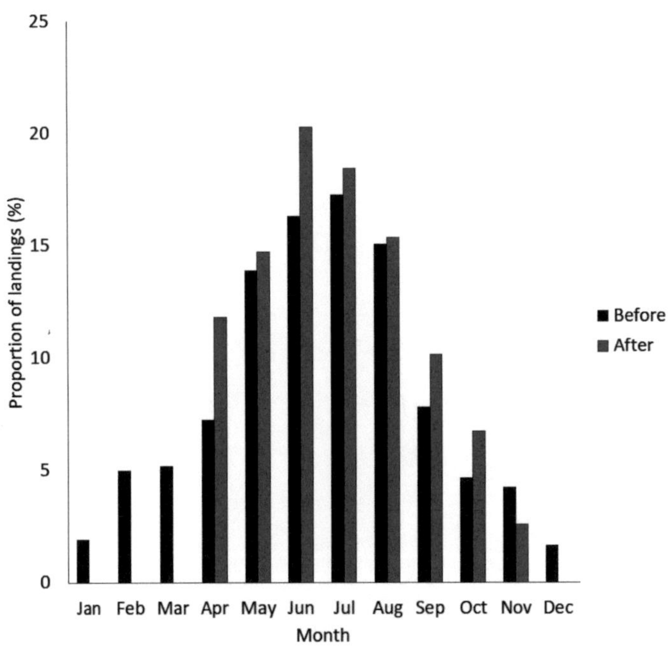

Figure 9.
Proportion of landings of limpets per month before (1990–2007) and after (2008–2017) the implementation of management measures in the archipelago of Madeira.

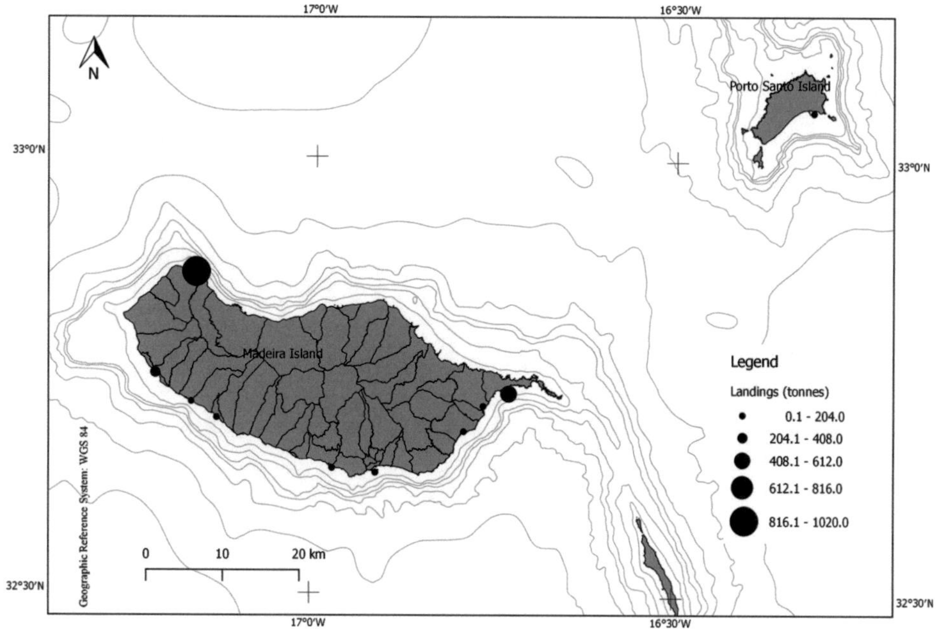

Figure 10.
Total landings of limpets per port between 1990 and 2017.

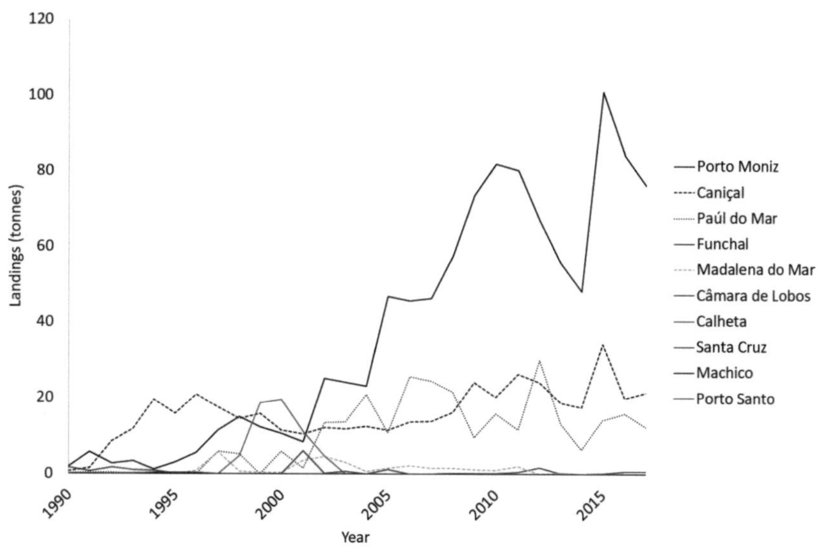

Figure 11.
Landings of limpets per year considering the landing port.

86% of the total of limpets landed in 2017. Since 2014, landings of commercially harvested limpets occur only in four ports (Porto Moniz, Caniçal, Paúl do Mar, and Funchal), mainly due to the proximity of the ports to the harvesting zones.

The commercial landings of limpets (in weight) in the archipelago of Madeira increased from 12% in 1990 to 96% of the total of molluscs landed in this region (**Figure 12**). Since 1998, limpets represent >90% of the landed molluscs in weight, increasing in importance over the years.

The economic importance of limpets in the fisheries sector of the archipelago of Madeira gradually increased over the years. In 1990, limpets represented

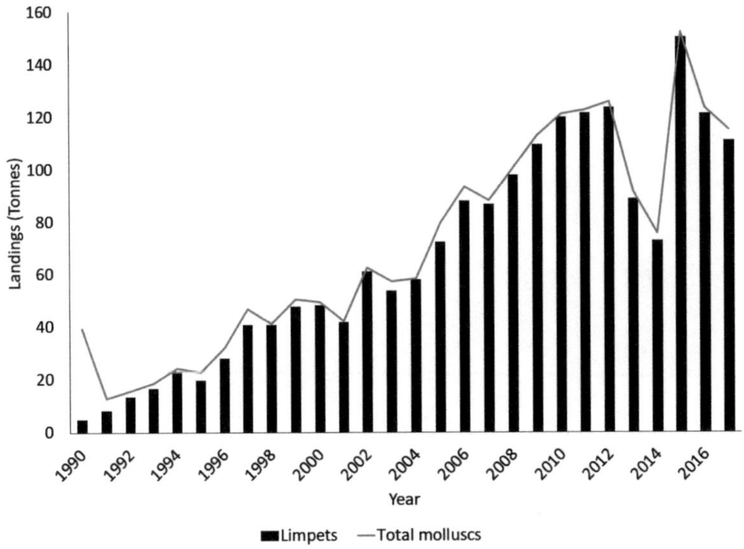

Figure 12.
Landings of limpets and the landings of the total molluscs per year, between 1990 and 2017 in the archipelago of Madeira.

approximately 26% of the total economic value landed for molluscs and in 2017 approximately 96%. Since 1991, limpets represent the majority of the landed value for molluscs. From 1998 to 2017, this resource represents over 90% of the landings of molluscs (**Figure 13**).

The traditional harvesting of limpets carried out by the local populations is not represented in the landings since current harvesting regulation does not require first auction sale for this activity (<3 kg/day/person), thus making it difficult to quantify the real impact of the traditional harvest on the exploited populations. Also, illegal harvest is not considered on the total of landings. Nonetheless, it is

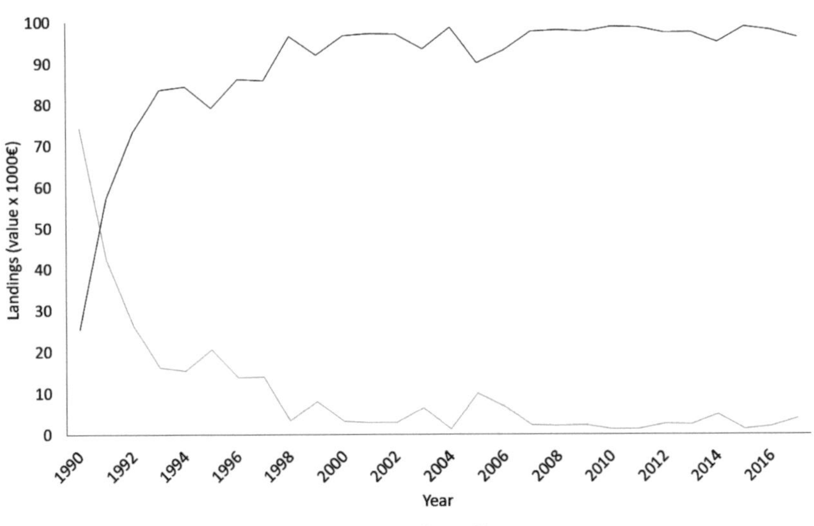

Figure 13.
Economic value of the landings of limpets and the economic value of the landings of the other molluscs per year, between 1990 and 2017 in the archipelago of Madeira.

known that commercial harvest occurs preferentially on certain locations (north coast of Madeira Island and around the Desertas), and, for these regions at least there is a better understanding of the harvesting effort on the exploited stocks.

The topshell (*P. sauciatus*) harvest in the archipelago of Madeira is carried out unregulated and without auction obligation, and as such, the landings of this species are residual, being impossible to estimate the real harvesting effort exerted on this species [2]. The landings varied from *ca.* 3 kg (2015) to *ca.* 230 kg (1993) between 1991 and 2017, yielding annual landed values between 14 and 520 € (**Figure 14**). Usually this species is sold directly to the markets at very high prices, reaching prices, i.e., between 15 and 20 € per kg.

4.2 Stock status

Specific stock assessments for limpets in the archipelago of Madeira begun in the mid-2000s and were the basis for regulation of shellfish harvest in the region [11]. Nevertheless, due to the increasing interest in these molluscs, preliminary monitoring of limpet populations began in 1996. More recently, studies on the biology and stock assessment of *P. candei* and *P. aspera* were accomplished by Henriques et al. [13] and Sousa et al. [14].

The stocks of *P. candei* and *P. aspera* were found to be underexploited but with exploitation occurring near their maximum sustainable yield in the archipelago of Madeira. However, limpets' specific life traits, like slow growth rates and long life-span, make these molluscs extremely vulnerable to overexploitation, and as such continuous monitoring of the stocks and the enforcement of the existing harvest regulations must be accomplished if future overexploitation is to be avoided [13, 14]. Surveillance and enforcement of the closed season are also particularly important to avoid poaching and safeguard the reproduction of these species. Specifically, for *P. aspera* since this species is a protandrous hermaphrodite, and thus especially vulnerable to size selective harvest, since after reaching sexual maturation a percentage of males shifts to females, thus the removal of larger individuals will target primarily females leading to a decreased reproductive success.

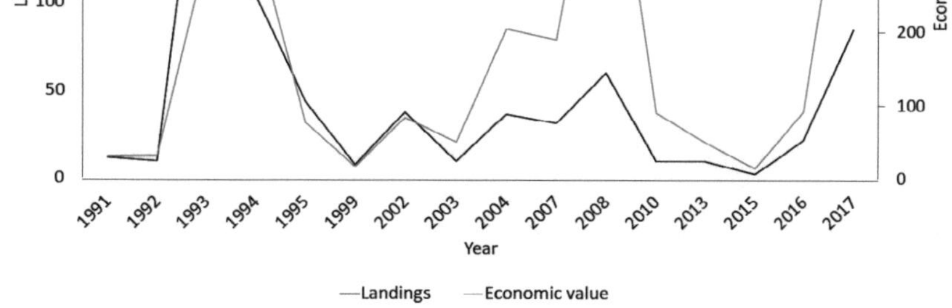

Figure 14.
Landings and economic value of topshells from 1990 to 2017 in the archipelago of Madeira.

Presently, some concerns persist regarding the management of the commercial exploitation of these two species in the region, namely, the lack of knowledge on the genetic connectivity between populations. In this sense, studies in this field are ongoing to establish proper conservation strategies considering connectivity or isolation of the populations. Another important concern relates to illegal poaching in the closed season and in MPAs and new conservation policies raising public awareness using innovative approaches involving not only decisionmakers but also the local communities which will be pivotal for the conservation of these species [12].

The first approach to assess the status of the stock of the topshell *P. sauciatus* in the archipelago of Madeira was held in 2018 although this species has been exploited since the fifteenth century. The stock of *P. sauciatus* seems to be moderately exploited in the region but vulnerable to the harvest of small specimens. Conservation measures on harvesting regulation, considering the biological and ecological specificities of this species in this region, are required to preserve the stock in the medium and long terms, and efforts in implementing it are currently underway.

5. Effects of management measures in limpet populations

The implementation of limpet harvest management measures in the archipelago of Madeira was of paramount importance due to the life history traits of these species allied to their economic importance that results in a high harvesting pressure on these resources.

The continuous monitoring of the limpet populations over time allowed the regulators to readjust harvesting regulation in Madeira, namely, the closed season period that initially lasted from November to January and presently ranges from December to March, to protect more effectively the breeding period of limpet species. This continuous monitoring clearly maximizes the returns that can be achieved through the regulation of this activity; since the responses of exploited stocks to harvest are continually changing, regulation has to adapt to these changes in order to promote its sustainability. The reduction of the number of harvesting licenses, the prohibition to capture immature individuals (<40 mm shell length), and the catch limits have also contributed to the increase of the mean size of the specimens, to the recovery of the size composition and to the increase of the abundance of the limpets' populations, in the archipelago of Madeira [3].

Conservation measures prompted an overall improvement of the exploited stocks on both exploited limpet species in Madeira. Comparative studies on the effectiveness of the implemented management measures demonstrated that the stocks of *P. aspera* and *P. candei* are slightly recovering since regulatory measures entered into force. The harvesting regulation was conducted to an increase in limpets mean shell length and to a more balanced size structure of the populations and a dominance of reproductive individuals. This effect was more noticeable in *P. aspera* populations due to the greater exploitation pressure exerted on this species [3].

The implemented management measures and the current levels of enforcement in the archipelago of Madeira showed positive results contrarily to those in the archipelago of Azores and the Canaries where management and enforcement were insufficient to protect the exploited limpet populations [15, 16].

The management measures led to an increase of 14% in proportion of reproductive individuals and to an increase in the size and age of first maturity for both species. A shift from a biased sex ratio before harvesting regulation to a balanced sex ratio after regulation occurred for *P. aspera* (R.S., pers. obs.).

Marine protected areas also contributed to the protection of limpet populations in the archipelago of Madeira. The increase of size-at-first maturity, shell size, balanced size composition, and capture per unit effort (CPUE) is evident in MPAs when compared to exploited populations. Also in the oldest and well-enforced MPAs, a high representation of large adults and more balanced populations for both limpet species (R.S., per. Obs.) was found. In fact, MPAs play a pivotal role in the recovery of the exploited stocks of limpets in the archipelago of Madeira, considering their potential to promote replenishment and recruitment in nearby coastal areas where limpets are or have been heavily exploited. However, further studies are required to confirm this possible positive effect of MPAs on limpet stocks.

6. Conclusions

The harvesting of limpets in the archipelago of Madeira is protected by legislation that provides the basis for a sustainable exploitation. However, it is still very common for violations to the current management policies and regulations which are punishable with the application of penalties when detected by the local authorities to occur. The two major weaknesses in this regard are the lack of compliance by the fishermen, which is mostly due to the lack of knowledge about the importance of the implemented management measures, and the poor enforcement by the regional authorities. A greater effort in the enforcement of regulations is required to ensure compliance of the fishing communities and an increase of the surveillance by the authorities to discourage illegal harvesting of these molluscs. Concerning topshells it is crucial to establish management measures on the harvesting of this resource to promote a sustainable exploitation in a medium and long term.

The ecological role that these key resources play in the coastal ecosystem and the importance of their survival not only from a conservational perspective but also economical needs to clearly transmitted to the fishermen and the general public, stakeholders, and authorities. This can be achieved through a proximity approach promoting training and involvement of all interested parties in the management and protection of these species. Promoting the awareness and advice to consumers in order to reduce illegal harvesting. Also, by instilling entrepreneurial spirit in fishermen by complementing their fishing activity with tourism through vessel trips and dives, adding value to these resources and increasing the fishermen income.

The implementation of obligatory species-specific landings for limpets is urgent to more accurately quantify the landings of each species and monitor the exploited stocks, since they are landed together as a mixed exploitation.

The encouragement of aquaculture investments and post-harvest facilities will reduce fishing pressure on exploited gastropods in the archipelago of Madeira and open the possibility of stock replenishment by reintroduction of cultured individuals to their natural habitat. This would also positively contribute to the socioeconomic development of the region, not only through production but also through job creation.

Acknowledgements

The authors are grateful to the Fisheries Research Service (DSI) of the Regional Directorate of Fisheries of the Autonomous Region of Madeira for providing the information used in this work. The first author (RS) was supported by a grant from ARDITI OOM/2016/010 (M1420-01-0145-FEDER-000001-Observatório Oceânico da Madeira-OOM) and the second author (JV) by a grant from FCT (SFRH/BSAB/143056/2018). The present study has been also supported by the

UE European Regional Development Fund in the framework of the Projects MARISCOMAC (MAC/2.3d/097), MACAROFOOD (MAC/2.3d/015), the Regional Government of Madeira, and FCT through the strategic project UID/MAR/04292/2019 granted to MARE.

Author details

Ricardo Sousa[1,2*], Rodrigo Riera[3], Joana Vasconcelos[3,4,5], Lídia Gouveia[2], Ana Rita Pinto[2], João Delgado[2,6], Adriana Alves[2], José A. González[7], Mafalda Freitas[8] and Paulo Henriques[9]

1 OOM/ARDITI–Oceanic Observatory of Madeira, Regional Agency for the Development of Research, Technology and Innovation, Funchal, Portugal

2 DSI/DRP—Research Service of the Regional Fisheries Department, Funchal, Portugal

3 Departamento de Ecología, Facultad de Ciencias, Universidad Católica de la Santísima Concepción, Concepción, Chile

4 Secretaria Regional de Educação, Funchal, Portugal

5 MARE—Marine and Environmental Sciences Centre, Portugal

6 CIIMAR—Interdisciplinary Centre of Marine and Environmental Research, Porto, Portugal

7 Ecología Marina Aplicada y Pesquerías (i-UNAT), Universidad de Las Palmas de Gran Canaria, Las Palmas de Gran Canaria, España

8 Estação de Biologia Marinha do Funchal e Centro de Ciências do Mar e do Ambiente (MARE), Funchal, Portugal

9 University of Madeira, Funchal, Portugal

*Address all correspondence to: ricardo.sousa@oom.arditi.pt

References

[1] Silva FA, Menezes CA. Elucidário Madeirense—I Volume A-E. Funchal: Tipografia Esperança; 1921. p. 826

[2] Sousa R, Vasconcelos J, Delgado J, Riera R, González JA, Freitas M, et al. Filling biological information gaps of the marine topshell *Phorcus sauciatus* (Gastropoda: Trochidae) to ensure its sustainable exploitation. Journal of the Marine Biological Association of the United Kingdom. 2018:1-9. DOI: 10.1017/S0025315418001054

[3] Sousa R, Vasconcelos J, Henriques P, Delgado J, Riera R. Long-term population status of two harvested intertidal grazers (*Patella aspera* and *Patella candei*), before (1996-2006) and after (2007-2017) the implementation of management measures. Journal of Sea Research. 2019;**144**:133-138. DOI: 10.1016/j.seares.2018.11.002

[4] Faria J, Pita A, Rivas M, Martins GM, Hawkins SJ, Ribeiro P, et al. A multiplex microsatellite tool for conservation genetics of the endemic limpet *Patella candei* in the Macaronesian archipelagos. Aquatic Conservation: Marine and Freshwater Ecosystems. 2016;**26**:775-781. DOI: 10.1002/aqc.2651

[5] Donald KM, Preston J, Williams ST, Reid DG, Winter D, Álvarez R, et al. Phylogenetic relationships elucidate colonization patterns in the intertidal grazers *Osilinus* Philippi, 1847 and *Phorcus* Risso, 1826 (Gastropoda: Trochidae) in the northeastern Atlantic Ocean and Mediterranean Sea. Molecular Phylogenetics and Evolution. 2012;**62**:35-45. DOI: 10.1016/j.ympev.2011.09.002

[6] Rico V, Santana JI, González JA. Técnicas de pesca artesanal en la Isla de Gran Canaria. Las Palmas: Monografías del Instituto Canario de Ciencias Marinas; 1999. p. 318

[7] Fernandes MJP. Riscos no Concelho da Ribeira Brava movimentos de vertente cheias rápidas e inundações [thesis]. Coimbra: Universidade de Coimbra; 2009

[8] Geldmacher J, Bogaard P, Hoernle K, Schmincke HU. The 40Ar/39Ar age dating of the Madeira archipelago and hotspot track (eastern North Atlantic). Geochemistry, Geophysics, Geosystems. 2000;**1**(2):1008-1034. DOI: 10.1029/1999GC000018

[9] Mitchell-Thomé RC. Radiometric studies in Macaronesia. Boletim do Museu Municipal do Funchal. 1995;**37**(167):52-85

[10] Alabsi N, Komatsu T. Characterization of fisheries management in Yemen: A case study of a developing country's management regime. Marine Policy. 2014;**50**:89-95. DOI: 10.1016/j.marpol.2014.05.015

[11] Delgado J, Alves A, Góis AR, Faria G, Henriques P, Correia J, et al. Exploração Comercial de Lapas na Madeira: Estudo Biológico e Contributo para a Gestão do Recurso. (Relatórios DBPO 01/2005). Funchal: Direção Regional de Pescas; 2005

[12] Henriques P, Delgado J, Sousa R. Patellid limpets: An overview of the biology and conservation of keystone species of the rocky shores. In: Ray S, editor. Organismal and Molecular Malacology. 1st ed. Croatia: Intech; 2017. pp. 71-95. DOI: 10.5772/67862

[13] Henriques P, Sousa R, Pinto AR, Delgado J, Faria G, Alves A, et al. Life history traits of the exploited limpet *Patella candei* (Mollusca: Patellogastropoda) of the North-Eastern Atlantic. Journal of the Marine Biological Association of the United Kingdom. 2012;**92**(6):1-9. DOI: 10.1017/S0025315411001068

[14] Sousa R, Delgado J, Pinto AR, Henriques P. Growth and reproduction of the northeastern Atlantic keystone species *Patella aspera* (Mollusca: Patellogastropoda). Helgoland Marine Research. 2017;**71**(8):1-13. DOI: 10.1186/s10152-017-0488-9

[15] Martins GM, Jenkins SR, Hawkins SJ, Neto AI, Medeiros AR, Thompson RC. Illegal harvesting affects the success of fishing closure areas. Journal of the Marine Biological Association of the United Kingdom. 2011;**91**(4):929-937. DOI: 10.1017/S0025315410001189

[16] Riera R, Pérez O, Álvarez O, Simón D, Díaz D, Monterroso O, et al. Clear regression of harvested intertidal mollusks. A 20-year (1994-2014) comparative study. Marine Environmental Research. 2016;**113**:56-61. DOI: 10.1016/j.marenvres.2015.11.003

Chapter 10

Assemblage of Gastropods in the Rocky Intertidal Zone of Asry Beach, Kingdom of Bahrain

Belen T. Lumeran

Abstract

The macrofaunal components find its habitat in all areas of the marine ecosystem. Specifically, gastropods are considered the most common inhabitants of the intertidal zone with wide range of distribution in the rocky intertidal biota. A 3-year study on gastropod assemblage in Asry Beach, Kingdom of Bahrain was conducted from 2016 to 2018 which determines diversity and evenness of the rocky intertidal species. Taxonomic identification showed 31 species which belong to 25 genera and 16 families. The total annual population of gastropod assemblage does not vary significantly at $p < 0.05$ using analysis of variance (ANOVA). Shannon-Wiener species diversity index (H′) revealed high species diversity (H′ = 4.19) in the *Family Muricidae*; moderate (H′ = 2.04–3.29) among the species of the *Family Pyramidellidae*, *Cerithiidae*, *Oliveliidae*, *Calliostomatidae*, *Turbinidae*, *Trochidae*, *Buccinidae*, *Velutinidae* and *Margeliidae*; and low (H′ = 1.04–1.11) among the species of the *Family Batillariidae*, *Borsoniidae*, *Cerethiopsidae*, *Epitoniidae*, *Volutidae* and *Columbellidae*. Pielou's evenness index show that most of the species have complete evenness (J' = 1.0–3.81). The varying annual mean temperature exert no effect on the total assemblage of gastropods (r = −0.0231). Tolerant species include *C. selectum*, *L. attenuatum* and *Turbonilla* sp. 1. The rocky intertidal pool is typically diverse although exposed to varying environmental occurrences.

Keywords: gastropod assemblage, rocky intertidal, tolerant species, Shannon-Wiener index, Pielou's evenness index

1. Introduction

The marine ecosystem is ecologically rich in terms of community of organisms which comprise the wide coastal area extending to a vast oceanic environment. The intertidal habitat had been the focus of many studies for ecological interactions which has an impact on diverse assemblage of macroflora and macrofauna. The rocky shore as an intertidal area is found at the shoreline between low and high tides and predominantly composed of solid rocks. Macrofaunal community tends to survive despite fluctuating daily diurnal tidal patterns and intense temperature [1]. This zone is considered highly productive with varying levels of biodiversity and dynamic due to diversity of both macrofauna and macroflora [2–4]. The wet rocky shore promotes enormous algal growth which serves as the feeding ground for the many macrofaunal community, specifically the gastropods thus promoting more diversity [5]. The rocky intertidal as a community is influenced by competitive processes primarily competition for space. The so called competitive dominance is manifested through the use of space. Community organization depends on consumer-prey interaction, physical disturbances as factors and competition for space [6]. Gastropods are the largest and the most diverse faunal group in *Phylum Mollusca*. The gastropods are known to have a wide range of distribution. They can live in a variety of habitats and are adapted to varying environmental conditions as affected by daily fluctuating tidal patterns and duration of exposure to sunlit which determine diurnal temperatures. These one-shelled marine invertebrates considered as economically important are the major inhabitants of the intertidal marine ecosystem. Marine gastropod species are varied and abundant as an important source of food for higher consumers and contribute to coastal food chains [7, 8]. Mollusks in the rocky shore are highly important in maintaining dynamics in the shore and ecological balance in beaches in addition to their key role in the trophic chains and nutrient recirculation [9]. Further, gastropods have important role in structuring the intertidal assemblages and regulating the intertidal communities since they can respond to variation in microhabitats in a vertical gradient and change their behavior in response to environmental occurrences [10]. Gastropod assemblage is influenced by the characteristic substrate which the majority of the shores are rocks. Characteristically, gastropods show variation in sizes, color and other phenotypic diversity [11]. Gastropods are motile however they are comparatively slow in locomotion which prevents them from moving into and out of the intertidal zone over a relatively short period of tidal range [12]. Hence these molluscan descendants have low migration potential which determines habitat stability in the rocky intertidal zone.

Studies on gastropod assemblage and diversity were conducted in various intertidal habitats of the marine ecosystem. A biodiversity study of gastropod in the intertidal zone of fine sand and coral reef was conducted in January 2017 in Sombu Beach, Wakatobi, Indonesia using quadrant plot method. Ten transects along the beach were established composed of four plots. There were 40 plots used as sources of data for analysis. The diversity of gastropod community was determined using Shannon-Wiener index [H′], evenness index (E) and dominance index (D). Results showed 13 species of gastropods which belong to three genera since the variety of substrate is low which has an impact on food resource and habitat of the identified gastropods [13]. A baseline study was conducted between November 2016 and February 2017 on marine gastropods diversity and distribution in two intertidal rocky shores of Terengganu, Peninsular Malaysia. The intertidal area was categorized and divided into three zones: upper, middle and lower. A 40 m length transect composed of six 1 m^2 quadrats was laid at random perpendicular to shore. Results revealed a total of five subclasses of gastropods which belong to nine families and 28 species from upper to lower intertidal zones. Diversity indices based on the results of Shannon-Wiener index (H) and Pielou's evenness index (J') were compared with reference to the identified zones [14]. In another study, a spatial distribution of macroinvertebrates on intertidal rocky shores was conducted in Gorgona Island, Colombia of the Tropical Eastern Pacific. Qualitative data were determined using rapid ecological assessments while quadrat method for quantitative data. Species richness, abundance and diversity were determined using Shannon-Wiener H′ and Pielou J' for evenness. Results of the study revealed 121 species of macroinvertebrates. Mollusks were the most abundant in terms of species and individual count. Researchers concluded that environmental stressor, heterogeneity and stability are limiting factors on the spatial distribution of macroinvertebrates species in this particular area of intertidal rocky shores [15]. A biodiversity study of gastropoda was conducted in the coastal waters of Ambon Island, Indonesia to determine the correlation between the physico-chemical factors and the diversity of coastal waters gastropods. The physical and chemical factors included temperature, salinity, pH and dissolved oxygen (DO). Results showed a total of 65 species in the two established research stations which belong to 48 genera, 19 families and 7 orders. The station-to station averages of the physico-chemical parameters were also determined. Results further show a very high diversity in both stations. Correlation analysis revealed a significant positive correlation between the physical and chemical factors (temperature, salinity, pH and DO) and the diversity of gastropods in the coastal waters of Ambon Island, Indonesia [16].

The Kingdom of Bahrain is endowed with a marine biota as habitat of all life forms. A number of public beach is found in Bahrain however Dry Dock Beach or commonly called Asry Beach in Hidd, Muharraq is of great interest. The beach as it is known to the people is part of the dry dock shipyard in the area. As a public beach, the area is often threatened by unaccountable wastes which may also affect the macrocomponents of the marine ecosystem. Hence continuous monitoring of the macroinvertebrates specifically the gastropods in the intertidal ecosystem is being undertaken since 2011 [17]. The intertidal ecosystem in Asry Beach is predominantly rocky and sandy. The shoreline is characteristically wide during ebb tide while very narrow during the spring tide. The rocky substrate extreme variation in temperature may exert drastic effect on the gastropod community. Hence, this study was conducted to determine the current status of the assemblage of gastropod population in the rocky intertidal zone of Asry Beach from January 2106 to January 2018. Specifically to identify and classify taxonomically the species of gastropod assemblage from the sampled quadrates; find out significant differences in the annual species assemblage of gastropods; calculate diversity using Shannon-Wiener (H′) diversity index; find out species evenness using Pielou's evenness index; and relate the effect of the aerial temperature on the total species assemblage of gastropods.

2. Research methods

2.1 Study site

The study area is located in the northeastern part of Dry Dock Beach commonly known as Asry Beach in Hidd, Muharraq having geographic coordinates of 26.1957°N and 50.6623°E. The total area of the intertidal zone is 206.30 m stretch where people usually perform beach activities during low tide. Part of the area is a 73.05 m stretch which composed the rocky intertidal biota (**Figure 1**). Both macrofauna and macroflora assemblages abound in the area in spite of the various human activities and other natural environmental occurrences.

2.2 Sampling, identification and field collection

This particular research was undertaken from January 2016 to January 2018 in a small parcel of the rocky environment of Dry Dock Beach commonly called Asry Beach in Muharraq, Kingdom of Bahrain. Sampling was done every month during the low tide between 7:00 and 1:00 pm. Belt transect method [12–14] was utilized for the quantitative assessment of gastropods in the rocky shore. A-50 m belt transect made of 0.45 mm clear nylon beading wire and 12 mm corrugated round steel bar post was established parallel to the elevated rocky shoreline during the low tide. The transect line was divided into 25 1 m^2 quadratic plots [13, 14] made of size 18 twisted nylon cord Bead Smith Super-Lon (S-Lon) on both sides for a total of 50 quadrates. The area where the transect line was laid represented the rocky intertidal ecosystem of the 73.05 m stretch. Random sampling [14] was conducted

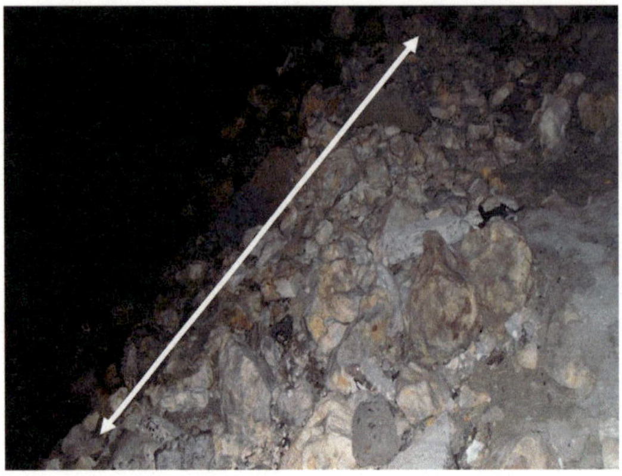

Figure 1.
Part of the 73.05 m stretch of the rocky shore during low tide (↔).

simultaneously for a total 25 sampled plots. Both live gastropod species and empty shells within the sampled quadrates including those in rock crevices were identified *in situ*. Aerial temperature was taken at the same time every sampling month using non-mercury thermometer.

Unidentified gastropod species and shells were photo-documented and collected. Specimens were placed in labeled transparent plastic bags. While in the field, the collected samples of live specimen were stored in an ice bucket to prevent desiccation. Collected samples were brought in the BioLab for identification to the lowest possible taxonomic level based on morphological characteristics using references in gastropods taxonomic identification [18–20]. The species names were also verified in a database [14] for marine species of gastropods [21]. Using taxonomic classification, identified gastropods were categorized into three hierarchy or categories namely: (1) *family*, (2) *genus*, and (3) *species*. Some live specimens in their natural habitat were photo-documented using Nikon D7000 camera. Collected specimens and shells were measured end-to-end from the longest point of axis for the length using Vernier caliper with ±0.01 mm accuracy and likewise photographed in the laboratory.

2.3 Data analysis

2.3.1 Determination of species diversity

Diversity of gastropod species was calculated using Shannon-Wiener Index (H′) with the formula: [13–15].

$$H' = -\sum_{i=}^{s} p_i \log p_i \tag{1}$$

where H′ = value of Shannon-Wiener diversity index; P_i = proportion of the ith species; $\log_e$ = natural logarithm of p_i; s = number of species in the community or species richness.

Using Shannon-Wiener diversity index (H′), species of gastropod in the assemblage is classified based on the following category: low (H′ < 2); moderate (2 < H′ < 4); and high (H′ > 4).

Species evenness index was determined using Pielou's evenness index with the formula:

$$J' = H'/H' max \tag{2}$$

where H′ = Shannon-Wiener diversity index; H_{max} = natural logarithm of species richness.

Species ranges from zero to one; zero means no evenness and one means complete evenness.

2.3.2 Statistical analysis

One-way analysis of variance (ANOVA) was used to find out significant differences in the total species assemblage. Correlation analysis using Pearson correlation coefficient *r* determined if the aerial temperature significantly affected the total annual species assemblage of gastropods.

3. Results and discussion

3.1 Study site

The characteristic substrate of the study site in Asry Beach is shown in **Figure 2**. It is an obligate, slightly elevated rocky platform which is often exposed to alternating tide sequences resulting to its periodic submergence and emergence. As part of the public beach, however people prefer the sandy part and spend their leisure time in water for swimming. Hence this area remains undisturbed which promotes massive growth of algae that provides nutrients and venue for ecological and intra-inter-specific interactions [5] among the faunal community of organisms. Based on observation and results of the study on the entire intertidal ecosystem [17] human factor is considerably not a limiting factor for the gastropod assemblage to be abundant in the area. This is in contrast to some research findings that human activities may intensify the exploitation of gastropod species for commercial purposes [11, 14]. Hence the rocky platform of the intertidal zone in this particular area serves as the habitation of the many life forms enduring the harsh environmental condition [1].

3.2 Sampling, identification and field collection

A total of 16 families composed of 25 genera and 31 species were identified in the rocky shore of Asry Beach from January 2016 to January 2018. The taxonomic

A B

Figure 2.
Cut view of the sampling site showing the rocky habitat (A) during low tide with massive algal growth (B).

identification, classification and total population of individual gastropod species in the sampled quadrates are presented in **Table 1**. Of the 16 families, *Family Muricidae* has four species which belong to four genera, *Nucella*, *Ocenebra*, *Ocinebrina* and *Scabrotrophon* while *Family Cerithiidae* [8] and *Oliveliidae* [7] has three species each. *Calliostoma selectum* of the *Family Calliostomatidae* has the most

Identified gastropods/Taxonomic hierarchy		**Year**			**Total**
		2016	2017	2018	
Family	**Species**				
Volutidae	*Arctomelon tenrsii*, Dall, 1872	3	3	6	**12**
Columbellidae	*Astyris aurantiaca*, Dall, 1871	1	2	1	**4**
Batillariidae	*Batillaria attramentaria*, Sowerby, 1855	3	3	3	**9**
Cerithiidae	*Bittium vancoverense*, Dall & Bartsch, 1910	4	4	7	**15**
	Lirobittium attenuatum, Carpenter, 1864	156	148	167	**471**
	Stylidium eschrichtii, Middendroft, 1849	10	17	21	**48**
Oliveliidae	*Callianax baetica*, Carpenter, 1864	5	7	9	**21**
	Callianax obiplicata, Sowerby, 1825	5	7	6	**18**
	Callianax pycna, S. S. Berry, 1925	2	3	7	**12**
Calliostomatidae	*Calliostoma selectum*, Dillwyn, 1817	1025	1221	1231	**3481**
	Calliostoma variegatum, Carpenter, 1864	54	68	81	**203**
Cerethiopsidae	*Cerithiopsis* sp.	63	71	77	**211**
Pyramidellidae	*Evalea tenuisculpta*, Carpenter, 1864	37	42	53	**132**
	Turbonilla sp. 1	68	77	81	**226**
	Turbonilla sp. 2	5	7	6	**18**
Turbinidae	*Homalopoma baculum*, Carpenter, 1864	28	38	42	**109**
	Homalopoma luridum, Dall, 1885	58	62	67	**187**
Buccidae	*Lirabuccinum odirum*, Reeve, 1846	6	8	11	**25**
	Volutharpa ampullarea, Middendroft, 1848	2	5	5	**12**
Trochidae	*Lirularia olirulata*, Carpenter, 1864	52	60	67	**179**
	Lirularia succinata, Carpenter, 1864	3	5	8	**16**
Velutinidae	*Marsenia thrombica*, Dall, 1871	9	15	9	**33**
	Velutina velutina, O. F. Muller, 1776	2	5	3	**10**
Muricidae	*Nucella lamellosa*, Gmelin, 1791	3	3	5	**11**
	Ocenebra inornata, Recluz, 1851	2	2	3	**7**
	Ocinebrina atropupurea, Carpenter, 1864	1	4	5	**10**
	Scabrotrophon maltzani, Kobett & Kuster, 1878	1	1	1	**3**
Epitoniidae	*Opalia borealis*, Kepe, 1881	7	10	13	**30**
Margeliidae	*Oepota olividensis*, Carpenter, 1864	1	1	3	**5**
	Oenopota tabulate, Carpenter, 1864	6	8	8	**22**
Borsoniidae	*Ophiodermelta cacellata*, Carpenter, 1864	1	1	1	**3**
Total		**1624**	**1908**	**2011**	**5543**
Total *Family*	**16**				
Total *Genus*	**25**				
Total *Species*	**31**				

Table 1.
Gastropod assemblage in the sampled quadrates using belt transect method.

number of the gastropods in the rocky intertidal zone having a total of 3841 followed by *Lirobittium attenuatum* of the *Family Cerithiidae* with 471, *Turbonilla* sp. 1 of the *Family Pyramidellidae* [7] with 226, *Cerithiopsis* sp., 211 of the *Family Cerethiopsidae*, and *Calliostoma variegatum*, 203 which is also from *Family Calliostomatidae*. Over-all, the highest number of total species was recorded in 2018 with 2011, a total of 1908 in 2017 while 1624 in 2016. In a 3-year study, a total of 5543 individual species were identified inhabiting the small parcel of the rocky shore of Asry Beach. Samples of gastropods in their natural habitat are shown in **Figure 3**, which include *Turbonilla* sp. (A) of the *Family Pyramidellidae*, *Lirularia olirulata* (B) of the *Family Trochidae*, and a group of *Cerithiopsis* sp. (C) of the *Family Cerethiopsidae* in association with algal community. Various studies on gastropod assemblage showed diversity of gastropods in the intertidal habitats [13–16] similar in this study with the use of diversity indices. However, the types of substrate vary from fine sand coral reef flats [13] and rocky shore beaches [9], including intertidal rocky shores [14, 15] and generally the intertidal zone of mixed substrate [6, 11]. Results of gastropod diversity study in Sombu Beach, Wakatobi, Indonesia showed 13 species of classes of gastropods covering only a month of study however using more transects [13] while a 3 year study of monthly sampling but using only one transect with 31 identified species. Generally, the type of substrate and the length of time for data gathering vary, although both conducted researches utilized the same sampling method. Composition of molluscs was compared in two communities of rocky shore beaches exposed to human activities. Results showed greater abundance and evenness in the site with less human activities [9] which is comparatively different from the claim that human factor considerably exerts no effect on diversity of gastropods in Asry Beach based on the results of monitoring studies [17]. Likewise, the researches being compared used diversity indices in quantitative analysis of data. Some similar researches covered wider spatial distribution of intertidal gastropods [1, 11, 14] although this research covered only a smaller scale on obligate rocky platform. Identified gastropods were taxonomically classified [13, 14] to the lowest level and specific count of species was determined.

3.3 Data analysis

3.3.1 Measure of diversity

Species diversity and evenness are presented in **Table 3**. Results of the Shannon-Wiener diversity index (H′) and Pielou's evenness index (J′) show that species diversity is high ($H' > 4$) in the *Family Muricidae* with computed H′ value of 4.19. Evenness is complete ($J' = 1$) where J′ value is 3.81. High species diversity of this group indicates the most number of species (four species) and evenly distributed in the sampled plots of the study site compared to other gastropod families. Generally, the muricids are commonly found in rocky habitat hence these are called rock snails. They are carnivores and predators of sessile animals [18, 19]. Identified gastropod assemblage can adapt to varying environmental conditions such as fluctuation in temperature due to intense solar radiation [1, 6] biotic and abiotic factors [11], substrate condition [10] and wave actions [1, 13]], physical and chemical factors such as temperature [1, 6], salinity, pH and dissolved oxygen (DO) [16] and increase human activities [9, 11]. Specifically in this particular research, aerial temperature was considered as a factor since transect with sampled plots were laid on the elevated rocky habitat (**Figure 1**). Recorded monthly aerial temperature from January 2016 to January 2018 was correlated with the population of gastropod species assemblage (**Table 1**). In some related gastropod diversity studies, temperature in general [16] was considered as limiting factor instead of aerial temperature.

Figure 3.
Samples of gastropods in their natural habitat during the lowest low tide.

In sampling locations partly submerged in water as in tidal pool [6, 12] coral reef habitat [13] and the littoral zone [14], water temperature was considered a moderating variable. Generally, the various habitats determined the species of gastropods [7] which directly or indirectly affected by the dynamics of the environmental temperatures. The distribution of organisms is non-homogeneous in the intertidal zone which changes based on the biotic and abiotic factors [12]. Species diversity is moderate ($2 < H' < 4$) among the *Family Pyramidellidae* ($H' = 3.29$; $J' = 2.99$); *Family Cerithiidae* ($H' = 3.22$; $J' = 2.92$); *Family Oliveliidae* ($H' = 3.12$; $J' = 2.84$); *Family Calliostomatidae* and *Turbinidae* ($H' = 2.19$; $J' = 1.99$) including species of *Family Trochidae*, *Buccidae*, *Velutinidae*, and *Margeliidae* having two to three species. Other species in the *Family Batillariidae*, *Borsoniidae*, *Cerethiopsidae*, *Epitoniidae*, *Volutidae* and *Columbellidae* have low diversity (H′ < 2). Of the 31 species (**Table 1**), three gastropod species have no evenness, *O. borealis* of the *Family Epitoniidae* (J' = 0.97) *A. ternsii* which belong to *Family Volutidae* and *A. aurantiaca of the Family Columbellidae* (J' = 0.95). Over-all, diversity in the rocky habitat of Asry Beach is moderate with the weighted mean of H′ = 2.06 however evenly distributed (J' = 1.88) (**Table 2**). Results imply that the resources specifically food [5] in the sampled rocky habitat supports the assemblage of gastropod species thus maintaining ecological balance in beaches [9].

Identified gastropods/Taxonomic classification		**Shannon-Wiener (H′)**	**Species evenness (J')**
Family	***Species***		
Volutidae	*Arctomelon ternsii*, Dall, 1872	1.04	0.95
Columbellidae	*Astyris aurantiaca*, Dall, 1871	1.04	0.95
Batillariidae	*Batillaria attramentaria* Sowerby, 1855	1.11	1.01
Cerithiidae	*Bittium vancoverense*, Dall & Bartsch, 1910*Lirobittium attenuatum*, Carpenter, 1864*Stylidium eschrichtii*, Middendroft, 1849	3.22	2.92
Oliveliidae	*Callianax baetica*, Carpenter, 1864*Callianax obiplicata*, Sowerby, 1825*Callianax pycna*, S. S. Berry, 1925	3.13	2.84
Calliostomatidae	*Calliostoma selectum*, Dillwyn, 1817*Calliostoma variegatum*, Carpenter, 1864	2.19	1.99
Cerethiopsidae	*Cerithiopsis* sp.	1.10	1.0
Pyramidellidae	*Evalea tenuisculpta*, Carpenter, 1864*Turbonilla* sp. 1*Turbonilla* sp. 2	3.29	2.99
Turbinidae	*Homalopoma baculum*, Carpenter, 1864*Homalopoma luridum*, Dall, 1885	2.19	1.99
Buccidae	*Lirabuccinum odirum*, Reeve, 1846*Volutharpa ampullarea*, Middendroft, 1848	2.10	1.91
Trochidae	*Lirularia olirulata*, Carpenter, 1864*Lirularia succinata*, Carpenter, 1864	2.12	1.93
Velutinidae	*Marsenia thrombica*, Dall, 1871*Velutina velutina*, O. F. Muller, 1776	2.08	1.89
Muricidae	*Nucella lamellosa*, Gmelin, 1791*Ocenebra inornata*, Recluz, 1851*Ocinebrina atropupurea*, Carpenter, 1864*Scabrotrophon maltzani*, Kobett & Kuster, 1878	4.19	3.81
Epitoniidae	*Opalia borealis*, Kepe, 1881	1.07	0.97
Margeliidae	*Oepota olividensis*, Carpenter, 1864*Oenopota tabulate*, Carpenter, 1864	2.04	1.85
Borsoniidae	*Ophiodermelta cacellata*, Carpenter, 1864	1.11	1.01
Weighted Mean		2.06	1.88

Table 2.
Result of Shannon-Wiener diversity index (H') and Pielou's species evenness (J').

3.3.2 Statistical analysis on the annual species assemblage

Although the total annual assemblage is high (**Table 1**) statistical analysis using one-way analysis of variance (ANOVA) revealed insignificant differences in the total annual population of gastropod species (**Table 3**). Results imply stability of gastropod species assemblage due to availability of food resources to support community structure [9]. Wet rocky biota promotes massive algal growth which serves as feeding ground for gastropod species [5]. Individual gastropod species only compete for space [6], be it on the rock surface or crevices. Although gastropods are motile, they tend to move slowly thus preventing them from moving in and out of the intertidal ecosystem as a consequence of short period of tidal change [12]. These attributes explain the stability of gastropod assemblage in this particular biota.

3.3.3 Effect of aerial temperature on the total assemblage of gastropods

Gastropods are adapted to various environmental factors [6, 13], including changes in diurnal temperature [1, 7, 8, 15]. Tolerant species are identified based on the highest annual individual species count and total count (**Table 1**). Top

Source	SS	df	MS	*F*-value	*p*-value
Between treatments	2653.5699	2	1326.7849	0.03052	.96995^{ns}
Within treatments	3912380.7097	90	43470.8968		
Total	3915034.2796	92			

ns = means not significant at p < 0.05.

Table 3.
Result of analysis of variance (ANOVA).

Month	Mean annual aerial temperature (in °C)			
	Year			Mean
	2016	2017	2018	
January	18	19	18	**18.33**
February	19	17	20	**18.67**
March	23	21	25	**23**
April	26	28	27	**27**
May	32	33	32	**32.33**
June	34	35	36	**35**
July	36	37	36	**36.33**
August	36	37	36	**36.33**
September	34	35	35	**34.67**
October	29	31	31	**30.33**
November	25	26	25	**25.33**
December	21	20	18	**19.67**
Mean	**27.75**	**28.25**	**28.25**	**28.08**

Table 4.
Summary of mean aerial temperature (in °C), January 2016–January 2018.

Variables	Sum	Mean	SS	n	r	p-value[ns]
Aerial temperature (°C)	333.99	27.832	508.163	12	−0.0231	.943438
Total species of gastropod assemblage	5543	461.917	11776022.917			

ns, not significant @ p < .05.

Table 5.
Result of Pearson r coefficient of correlation.

tolerant species include *C. selectum* with increasing annual population of 1025 in 2016, 1221 in 2017, and 2231 in 2018 (total of 3481); *L. attenuatum* (2016 = 156; 2017 = 148; and 2018 = 167; total = 471); *Turbonilla* sp. 1 (2016 = 68; 2017 = 71; and 2018 = 81; total of 226). Results imply that these species have higher level of tolerance to temperature changes. The mean temperatures vary annually. In 2016, the mean monthly temperature ranged from 18 to 36°C; 17 to 37°C in 2017; and 18 to 36°C in 2018. Hence, the mean temperature in **Table 4** shows fluctuating monthly temperatures with annual mean temperature of 27.75°C in 2016 and 28.25°C in 2017 and 2018.

Statistically, result of Pearson r coefficient of correlation in **Table 5** shows a negative correlation ($r = -0.0231$). It means that temperature changes exert no effect on the total count of gastropod assemblage. The p-value of .943438 means not significant at $p < .05$. The identified assemblage of gastropods in the rocky zone of Asry Beach constitutes a stable community structure which means that aerial temperature has insignificant effect (p-value = .943438, not significant at $p < .05$) on the population of gastropods although the rocky habitat is exposed to diurnal sun lit. In some related studies however [16], correlation analysis revealed a significant positive correlation between the physical factors such as temperature, salinity, pH and DO in diversity of gastropods in the coastal waters of Ambon Island, Indonesia. Temperature as a factor in the related study refers to water temperature considering the type of substrate where the gastropods were sampled. Hence different from the emergent rocky habitat where the gastropods abound on rock surfaces and crevices (**Figure 2**).

4. Conclusions

Asry Beach in the Kingdom of Bahrain is typically an intertidal ecosystem characterized by both sandy and rocky substrate. As a public beach, visitors and other beach goers prefer the sandy portion rather than the rocky part. The rocky zone remains undisturbed where macroflora/fauna community contributes in the coastal food chain and in other ecological interactions. The emergent wet rock surfaces and crevices promote massive algal growth which serves as the feeding ground for many macrofaunal communities. Marine gastropods as natural inhabitants in the rocky biota are tolerable to changes in aerial temperature, thus maintains the dynamics in the rocky shores and ecological balance in beaches. The over-all status of diversity and individual species distribution determine the impact of both the biotic and abiotic factors on gastropod assemblage. Periodic biodiversity assessment and monitoring are initiatives for the protection, preservation and conservation of the natural habitat of gastropods.

Author details

Belen T. Lumeran
Faculty, Department of Mathematics and Sciences, AMA International University-Bahrain, Kingdom of Bahrain

*Address all correspondence to: belentlumeran@gmail.com

References

[1] Dave TH, Chudasama BG. Survey and diversity of intertidal mollusks along the coast of Veraval (Gujarat), Arabian Sea. International Journal of Science, Environment and Technology. 2018;**7**:353-360

[2] Vaghela A, Bhadja P, Kundu R. Diversity and distribution of intertidal mollusca at Saurashtra coast of Arabian Sea, India. Global Journal of Bio-Science and Biotechnology. 2013;**2**:154-158

[3] Lakwal VR, Kharate DS, Mokashe SS. Intertidal macrofaunal invertebrates diversity from Ratnagiri coast, (MS) India of Arabian Sea. International Journal for Research in Applied Science and Engineering Technology. 2018;**6**:1630-1638

[4] Donnarumma L, Sadulli R, Appolloni L, Sanchez-Lizazo JL, Russo GP. Assessment of structural and functional diversity of mollusk assemblages within vermetid bioconstructions. Diversity. 2018;**10**:96. DOI: 10.3390/d10030096

[5] Jeeva C, Mohan PM, Dil Baseer Sabith KK, Ubare VV, Murugananthar M, Kumari RK. Distribution of gastropods in the intertidal environment of south, middle and north Andaman Islands, India. Open Journal of Marine Science. 2018;**8**:173-195. DOI: 10.4236/ojms.81009

[6] Sullivan H. The effect of gastropod diversity and density on macroalgae diversity in the Massachusetts rocky intertidal. Journal of Marine Ecology. 2015;**1**:19-26

[7] Kohan A, Badbardast Z, Shokri M. The gastropods fauna along the Bushehr Province intertidal zone of the Persian Gulf. Journal of the Persian Gulf. 2012;**3**:33-42

[8] Siti-Balkhis AB, Yaman IC, Siti-Hasmah I, Khalil MZ, Muhammad-Shukri MY, Zulfigar Y, et al. A survey of the marine intertidal macrogastropoda in the northern straits of Malaca. ASM Science Journal. 2015;**8**(2):159-164

[9] Martinez E, Sanchez J, Alba A, Vazquez AA. Changes in structure and composition of two communities of rocky shores mollusks exposed to different human impact in playa Jibacoa, Cuba. Journal of Ecosystem & Ecography. 2015;**5**:1-6. DOI: 10.4172/2157-7625.1000162

[10] Pandey V, Thiruchitrambalam G, Satyam K. Habitat heterogeneity determines structural properties of intertidal gastropod assemblages in a pristine tropical island ecosystem. Indian Journal of Geo Marine Sciences. 2018;**47**:846-853

[11] Kurhe AR, Rodriguez MA, Suryawanshi GD. Vertical distribution and diversity of gastropods mollusks from intertidal habitats of the Ratnagiri coast Maharashtra, India. International Research Journal of Natural and Applied Sciences. 2014;**1**:1-13

[12] Pandey M, Desai AY, Mathew KL. Quantitative abundance of key intertidal gastropods at port Okha reef Gujarat. International Journal of Fisheries and Aquatic Studies. 2017;**5**:188-192

[13] Rahmayanti F, Nazira FK, Dewi AK, Tranggono YA. Biodiversity of gastropod in the Sombo Beach, Wakatobi, Indonesia. In: IOP Conference Series; 2018. DOI: 10.1088/1755-1315

[14] Baharuddin N, Basri NB, Syawal NH. Marine gastropods (Gastropoda; Mollusca) diversity and distribution on intertidal rocky shores of Terengganu, peninsular Malaysia. AACL Bioflux. 2018;**11**:1144-1154

[15] Londono-Cruz E, Lopez de Mesa-Agudelo LA, Arias-Galvez F, Herrera-Paz DL, Prado A, Cuellar LM, et al. Distribution of macroinvertebrates on intertidal rocky shores in Gorgona Island, Colombia (tropical eastern Pacific). Journal of Tropical Biology. 2014;**62**:189-198

[16] Rumahlatu D, Leiwakabessy F. Biodiversity of gastropods in the coastal waters of Ambon Island, Indonesia. AACL Bioflux. 2017;**10**:1-12

[17] Lumeran BT. Gastropods in the intertidal zone of Asry Beach, Kingdom of Bahrain. International Advanced Research Journal in Science, Engineering and Technology. 2015;**2**:93-97. DOI: 10.17148/IARJSET.2015.2720

[18] Harbo RM. A Field Guide to Seashells and Shellfish of the Pacific Northwest. Canada: Harbour Publishing Company Limited; 2009. 22 pp

[19] Robin A. Encyclopedia of Marine Gastropods. Germany: ConchBooks, IKAN Interwasser-Archi; 2008. 480 pp

[20] Shell Catalogue [Internet]. 2019. Available from: www.gastropods.com/taxon_pages/TN_Family-shtml

[21] World Register of Marine Species-Gastropods (WoRMS) [Internet]. 2019. Available from: www.marinespecies.org/